MAN'S PLACE IN

THE UNIVERSE

A Study of the Results of Scientific Research
in Relation to the Unity or Plurality
of Worlds

BY

ALFRED R. WALLACE

LL.D., D.C.L., F.R.S., ETC.

'O, glittering host! O, golden line! I would I had an
angel's ken, Your deepest secrets to divine, And read your
mysteries to men.'

THIRD EDITION

1904

British Library Cataloguing-in-Publication Data
A catalogue record for this book is available from the
British Library

Alfred Russel Wallace

Alfred Russel Wallace was born on 8th January 1823 in the village of Llanbadoc, in Monmouthshire, Wales.

At the age of five, Wallace's family moved to Hertford where he later enrolled at Hertford Grammar School. He was educated there until financial difficulties forced his family to withdraw him in 1836. He then boarded with his older brother John before becoming an apprentice to his eldest brother, William, a surveyor. He worked for William for six years until the business declined due to difficult economic conditions.

After a brief period of unemployment, he was hired as a master at the Collegiate School in Leicester to teach drawing, map-making, and surveying. During this time he met the entomologist Henry Bates who inspired Wallace to begin collecting insects. He and bates continued exchanging letters after Wallace left teaching to pursue his surveying career. They corresponded on prominent works of the time such as Charles Darwin's *The Voyage of the Beagle* (1839) and Robert Chamber's *Vestiges of the Natural History of Creation* (1844).

Wallace was inspired by the travelling naturalists of the day and decided to begin his exploration career collecting specimens in the Amazon rainforest. He explored the Rio Negra for four years, making notes on the peoples and languages he encountered as well as the geography, flora, and fauna. On his return voyage his ship, Helen, caught fire and he and the crew were stranded for ten days before being picked up by the Jordeson, a brig travelling from Cuba to London. All of his specimens aboard Helen had been lost.

After a brief stay in England he embarked on a journey to the Malay Archipelago (now Singapore, Malaysia, and Indonesia). During this eight year period he collected more than 126,000 specimens, several thousand of which represented new species to science. While travelling, Wallace refined his thoughts about evolution and in 1858 he outlined his theory of natural selection in an article he sent to Charles Darwin. This was published in the same year along with Darwin's own theory. Wallace eventually published an account of his travels *The Malay Archipelago* in 1869, and it became one of the most popular books of scientific exploration in the 19th century.

Upon his return to England, in 1862, Wallace became a staunch defender of Darwin's landmark work *On the Origin of Species* (1859). He wrote responses to those critical of the theory of natural selection, including 'Remarks on the Rev. S. Haughton's Paper on the Bee's Cell, And on the Origin of Species' (1863) and 'Creation by Law' (1867). The former of these was particularly pleasing to Darwin. Wallace also published important papers such as 'The Origin of Human Races and the Antiquity of Man Deduced from the Theory of 'Natural Selection" (1864) and books, including the much cited *Darwinism* (1889).

Wallace made a huge contribution to the natural sciences and he will continue to be remembered as one of the key figures in the development of evolutionary theory.

Wallace died on 7th November 1913 at the age of 90. He is buried in a small cemetery at Broadstone, Dorset, England.

MAN'S PLACE IN
THE UNIVERSE

'I said unto my inmost heart, Shall I don corslet, helm, and shield, And shall I with a Giant strive, And charge a Dragon on the field?'

J.H. Dell.

PREFACE

This work has been written in consequence of the[v] great interest excited by my article, under the same title, which appeared simultaneously in *The Fortnightly Review* and the *New York Independent*. Two friends who read the manuscript were of opinion that a volume, in which the evidence could be given much more fully, would be desirable, and the result of the publication of the article confirmed their view.

I was led to a study of the subject when writing four new chapters on Astronomy for a new edition of *The Wonderful Century*. I then found that almost all writers on general astronomy, from Sir John Herschel to Professor Simon Newcomb and Sir Norman Lockyer, stated, as an indisputable fact, that our sun is situated *in* the plane of the great ring of the Milky Way, and also very nearly in the centre of that ring. The most recent researches also showed that there was little or no proof of there being any stars or nebulæ very far beyond the Milky Way, which thus seemed to be the limit, in that direction, of the stellar universe.[vi]

Turning to the earth and the other planets of the Solar System, I found that the most recent researches led to the conclusion that no other planet was likely to be the seat of organic life, unless perhaps of a very low type. For many years I had paid special attention to the problem of the measurement of geological time, and also that of the mild climates and generally uniform conditions that had prevailed throughout all geological epochs; and on considering the number of concurrent causes and the delicate balance of conditions required to maintain such uniformity, I became still more convinced that the evidence was exceedingly strong against the probability or possibility of any other planet being inhabited.

Having long been acquainted with most of the works dealing with the question of the supposed *Plurality of Worlds*, I was quite aware of the very superficial treatment the subject had received, even in the hands of the most able writers, and this made me the more willing to set forth the whole of the available evidence—astronomical, physical, and biological—in such a way as to show both what was proved and what suggested by it.

The present work is the result, and I venture to think that those who will read it carefully will admit that it is a book that was worth writing. It is founded almost entirely on the marvellous body of facts and[vii] conclusions of the New Astronomy together with those reached by modern physicists, chemists, and biologists. Its novelty consists in combining the various results of these different branches of science into a connected whole, so as to show their bearing upon a single problem—a problem which is of very great interest to ourselves.

This problem is, whether or no the logical inferences to be drawn from the various results of modern science lend support to the view that our earth is the only inhabited planet, not only in the Solar System but in the whole stellar universe. Of course it is a point as to which absolute demonstration, one way or the other, is impossible. But in the absence of any direct proofs, it is clearly rational to inquire into probabilities; and these probabilities must be determined not by our prepossessions for any particular view, but by an absolutely impartial and unprejudiced examination of the tendency of the evidence.

As the book is written for the general, educated body of readers, many of whom may not be acquainted with any aspect of the subject or with the wonderful advance of recent knowledge in that department often termed the New Astronomy, a popular account has been given of all those branches of it which bear upon the special subject here discussed. This part of the work occupies the first[viii] six chapters. Those who are fairly acquainted with modern astronomical literature, as given in popular works, may begin at my seventh chapter, which marks the commencement of the considerable body of evidence and of argument I have been able to adduce.

To those of my readers who may have been influenced by any of the adverse criticisms on my views as set forth in the article already referred to, I must again urge, that throughout the whole of this work, neither the facts nor the more obvious conclusions from the facts are given on my own authority, but always on that of the best astronomers, mathematicians, and other men of science to whose works I have had access, and whose names, with exact references, I generally give.

What I claim to have done is, to have brought together the various facts and phenomena *they* have accumulated; to have set forth the hypotheses by which *they* account for them, or the results to which the evidence clearly points; to have judged between conflicting opinions and theories; and lastly, to have combined the results of the various widely-separated departments of science, and to have shown how they bear upon the great problem which I have here endeavoured, in some slight degree, to elucidate.

As such a large body of facts and arguments from[ix] distinct sciences have been here brought together, I have given a rather full summary of the whole argument, and have stated my final conclusions in six short sentences. I then briefly discuss the two aspects of the whole problem—those from the materialistic and from the spiritualistic points of view; and I conclude with a few general observations on the almost unthinkable problems raised by ideas of Infinity— problems which some of my critics thought I had attempted in some degree to deal with, but which, I here point out, are altogether above and beyond the questions I have discussed, and equally above and beyond the highest powers of the human intellect.

Broadstone, Dorset,
September 1903.

'The wilder'd mind is tost and lost, O sea, in thy eternal tide; The reeling brain essays in vain, O stars, to grasp the vastness wide! The terrible tremendous scheme That glimmers in each glancing light, O night, O stars, too rudely jars The finite with the infinite!'

J.H. Dell.

CONTENTS

CONTENTS

CONTENTS

EIGHT DIAGRAMS IN THE TEXT AND

TWO STAR CHARTS AT END.

'Who is man, and what his place? Anxious asks the heart, perplext In this recklessness of space, Worlds with worlds thus intermixt: What has he, this atom creature, In the infinitude of Nature?'

F.T. Palgrave.

MAN'S PLACE IN THE UNIVERSE

CHAPTER I

EARLY IDEAS AS TO THE UNIVERSE AND ITS RELATION TO MAN

When men attained to sufficient intelligence for speculations as to their own nature and that of the earth on which they lived, they must have been profoundly impressed by the nightly pageant of the starry heavens. The intense sparkling brilliancy of Sirius and Vega, the more massive and steady luminosity of Jupiter and Venus, the strange grouping of the brighter stars into constellations to which fantastic names indicating their resemblance to various animals or terrestrial objects seemed appropriate and were soon generally adopted, together with the apparently innumerable stars of less and less brilliancy scattered broadcast over the sky, many only being visible on the clearest nights and to the acutest vision, constituted altogether a scene of marvellous and impressive splendour of which it must have seemed almost impossible to attain any real knowledge, but which afforded an endless field for the imagination of the observer.

The relation of the stars to the sun and moon in their respective motions was one of the earliest problems for the astronomer, and it was only solved by careful and continuous observation, which showed that the invisibility of the former during the day was wholly due to the blaze of light, and this is said to have been proved at an early period by the observed fact that from the bottom of very deep wells stars can be seen while the sun is shining. During total eclipses of the sun also the brighter stars become visible, and, taken in connection with the fixity of position of the pole-star, and the course of those circumpolar stars which never set in the latitudes of Greece, Egypt, and Chaldea, it soon became possible to frame a simple hypothesis which supposed the earth to be suspended in space, while at an unknown distance from it a crystal sphere revolved upon an axis indicated by the pole-star, and carried with it the whole host of heavenly bodies. This was the theory of Anaximander (540 B.C.), and it served as the starting-point for the more complex theory which continued to be held in various forms and with endless modifications down to the end of the sixteenth century.

It is believed that the early Greeks obtained some knowledge of astronomy from the Chaldeans, who appear to have been the first systematic observers of the heavenly bodies by means of instruments, and who are said to have discovered the cycle of eighteen years and ten days after which the sun and moon return to the same relative positions as seen from the earth. The Egyptians perhaps derived their knowledge from the same source, but there is no proof that they were great observers, and the accurate orientation, proportions, and angles of the Great Pyramid and its inner passages may perhaps indicate a Chaldean architect.

The very obvious dependence of the whole life of the earth upon the sun, as a giver of heat and light, sufficiently explains the origin of the belief that the latter was a mere appanage of the former; and as the moon also illuminates the night, while the stars as a whole also give a very perceptible amount of light, especially in the dry climate and clear atmosphere of the East, and when compared with the pitchy darkness of cloudy nights when the moon is below the horizon, it seemed clear that the whole of these grand luminaries—sun, moon, stars, and planets—were but parts of the terrestrial system, and existed solely for the benefit of its inhabitants.

Empedocles (444 B.C.) is said to have been the first who separated the planets from the fixed stars, by observing their very peculiar motions, while Pythagoras and his followers determined correctly the order of their succession from Mercury to Saturn. No attempt was made to explain these motions till a century later, when Eudoxus of Cnidos, a contemporary of Plato and of Aristotle, resided for some time in Egypt, where he became a skilful astronomer. He was the first who systematically worked out and explained the various motions of the heavenly bodies on the theory of circular and uniform motion round the earth as a centre, by means of a series of concentric spheres, each revolving at a different rate and on a different axis, but so united that all shared in the motion round the polar axis. The moon, for example, was supposed to be carried by three spheres, the first revolved parallel to the equator and accounted for the diurnal motion—the rising and setting—of the moon; another moved parallel to the ecliptic and explained the monthly changes of the moon; while the third revolved at the same rate but more obliquely, and explained the inclination of the moon's orbit to that of the earth. In the same way each of the five planets had four spheres, two moving like the first two of the moon, another one also moving in the ecliptic was required to explain the retrograde motion of the planets, while a fourth oblique to the ecliptic was needed to explain the diverging motions due to the different obliquity of the orbit of each planet to that of the earth. This was the celebrated Ptolemaic system in the simplest form needed to account for the more obvious motions of the heavenly bodies. But in the course of ages the Greek and Arabian astronomical observers discovered small divergences due to the various degrees of excentricity of the orbits of the moon and planets and their consequent varying rates of motion; and to explain these other spheres were added, together with smaller circles sometimes revolving excentrically, so that at length about sixty of these spheres, epicycles and excentrics were required to account for the various motions observed with the rude instruments, and the rates of motion determined by the very imperfect time-measurers of those early ages. And although a few great philosophers had at different times rejected this cumbrous system and had endeavoured to promulgate more correct ideas, their views had no influence on public opinion even among astronomers and mathematicians, and the Ptolemaic system held full sway down to the time of Copernicus, and was not finally given up till Kepler's *Laws* and Galileo's *Dialogues* compelled the adoption of simpler and more intelligible theories.

We are now so accustomed to look upon the main facts of astronomy as mere elementary knowledge that it is difficult for us to picture to ourselves the state of almost complete ignorance which prevailed even among the most civilised nations throughout antiquity and the Middle Ages. The rotundity of the earth was held by a few at a very early period, and was fairly well established in later classical times. The rough determination of the size of our globe followed soon after; and when instrumental observations became more perfect, the distance and size of the moon were measured with sufficient accuracy to show that it was very much smaller than the earth. But this was the farthest limit of the determination of astronomical sizes and distances before the discovery of the telescope. Of the sun's real distance and size nothing was known except that it was much farther from us and much larger than the moon; but even in the century before the commencement of the Christian era Posidonius determined the circumference of the earth to be 240,000 stadia, equal to about 28,600 miles, a wonderfully close approximation considering the very imperfect data at his command. He is also said to have calculated the sun's distance, making it only one-third less than the true amount, but this must have been a chance coincidence, since he had no means of measuring angles more accurately than to one degree, whereas in the determination of the sun's distance instruments are required which measure to a second of arc.

Before the discovery of the telescope the sizes of the planets were quite unknown, while the most that could be ascertained about the stars was, that they were at a very great distance from us. This being the extent of the knowledge of the ancients as to the actual dimensions and constitution of the visible universe, of which, be it remembered, the earth was held to be the centre, we cannot be surprised at the almost universal belief that this universe existed solely for the earth and its inhabitants. In classical times it was held to be at once the dwelling-place of the gods and their gift to man, while in Christian ages this belief was but slightly, if at all, changed; and in both it would have been considered impious to maintain that the planets and stars did not exist for the service and delight of mankind alone but in all probability had their own inhabitants, who might in some cases be even superior in intellect to man himself. But apparently, during the whole period of which we are now treating, no one was so daring as even to suggest that there were other worlds with other inhabitants, and it was no doubt because of the idea that we occupied *the* world, the very centre of the whole surrounding universe which existed solely for us, that the discoveries of Copernicus, Tycho Brahé, Kepler, and Galileo excited so much antagonism and were held to be impious and altogether incredible. They seemed to upset the whole accepted order of nature, and to degrade man by removing his dwelling-place, the earth, from the commanding central position it had always before occupied.

CHAPTER II

MODERN IDEAS AS TO MAN'S RELATION TO THE UNIVERSE

The beliefs as to the subordinate position held by sun, moon, and stars in relation to the earth, which were almost universal down to the time of Copernicus, began to give way when the discoveries of Kepler and the revelations of the telescope demonstrated that our earth was not specially distinguished from the other planets by any superiority of size or position. The idea at once arose that the other planets might be inhabited; and when the rapidly increasing power of the telescope, and of astronomical instruments generally, revealed the wonders of the solar system and the ever-increasing numbers of the fixed stars, the belief in other inhabited worlds became as general as the opposite belief had been in all preceding ages, and it is still held in modified forms to the present day.

But it may be truly said that the later like the earlier belief is founded more upon religious ideas than upon a scientific and careful examination of the whole of the facts both astronomical, physical, and biological, and we must agree with the late Dr. Whewell, that the belief that other planets are inhabited has been generally entertained, not in consequence of physical reasons but in spite of them. And he adds:—'It was held that Venus, or that Saturn was inhabited, not because anyone could devise, with any degree of probability, any organised structure which would be suitable to animal existence on the surfaces of those planets; but because it was conceived that the greatness or goodness of the Creator, or His wisdom, or some other of His attributes, would be manifestly imperfect, if these planets were not tenanted by living creatures.' Those persons who have only heard that many eminent astronomers down to our own day have upheld the belief in a 'Plurality of Worlds' will naturally suppose that there must be some very cogent arguments in its favour, and that it must be supported by a considerable body of more or less conclusive facts. They will therefore probably be surprised to hear that any direct evidence which may be held to support the view is almost wholly wanting, and that the greater part of the arguments are weak and flimsy in the extreme.

Of late years, it is true, some few writers have ventured to point out how many difficulties there are in the way of accepting the belief, but even these have never examined the question from the various points of view which are essential to a proper consideration of it; while, so far as it is still upheld, it is thought sufficient to show, that in the case of some of the planets, there seem to be such conditions as to render life possible. In the millions of planetary systems supposed to exist it is held to be incredible that there are not great numbers as well fitted to be inhabited by animals of all grades, including some as high as man or even higher, and that we must, therefore, believe that they are so inhabited. As in the present work I propose to show, that the probabilities and the weight of direct evidence tend to an exactly opposite conclusion, it will be well to pass briefly in review the various writers on the subject, and to give some indication of the arguments they have used and the facts they have set forth. For the earlier upholders of the theory I am indebted to Dr. Whewell, who, in his *Dialogue on the Plurality of Worlds*—a Supplement to his well-known volume on the subject— refers to all writers of importance known to him.

The earliest are the great astronomers Kepler and Huygens, and the learned Bishop Wilkins, who all believed that the moon was or might probably be inhabited; and of these Whewell considers Wilkins to have been by far the most thoughtful and earnest in supporting his views. Then we have Sir Isaac Newton himself who, at considerable length, argued that the sun was probably inhabited. But the first regular work devoted to the subject appears to have been written by M. Fontenelle, Secretary to the Academy of Sciences in Paris, who in 1686 published his *Conversations on the Plurality of Worlds*. The book consisted of five chapters, the first explaining the Copernican Theory; the second maintaining that the moon is a habitable world; the third gives particulars as to the moon, and argues that the other planets are also inhabited; the fourth gives details as to the worlds of the five planets; while the fifth declares that the fixed stars are suns, and that each illuminates a world. This work was so well written, and the subject proved so attractive, that it was translated into all the chief European languages, while the astronomer Lalande edited one of the French editions. Three English translations were published, and one of these went through six editions down to the year 1737. The influence of this work was very great and no doubt led to that general acceptance of the theory by such men as Sir William Herschel, Sir John Herschel, Dr. Chalmers, Dr. Dick, Dr. Isaac Taylor, and M. Arago, although it was wholly founded on pure speculation, and there was nothing that could be called evidence on one side or the other.

This was the state of public opinion when an anonymous work appeared (in 1853) under the somewhat misleading title of *The Plurality of Worlds: An Essay*. This was written, as already stated, by Dr. Whewell, who, for the first time, ventured to doubt the generally accepted theory, and showed that all the evidence at our command led to the conclusion that some of the planets were *certainly* not habitable, that others were *probably* not so, while in none was there that close correspondence with terrestrial conditions which seemed essential for their habitability by the higher animals or by man. The book was ably written and showed considerable knowledge of the science of the time, but it was very diffuse, and the larger part of it was devoted to showing that his views were not in any way opposed to religion. One of his best arguments was founded on the proposition that '*the Earth's Orbit is the Temperate Zone of the Solar System*,' that there only is it possible to have those moderate variations of heat and cold, dryness and moisture, which are suitable for animal life. He suggested that the outer planets of the system consisted mainly of water, gases, and vapour, as indicated by their low specific gravity, and were therefore quite unsuitable for terrestrial life; while those near the sun were equally unsuited, because, owing to the great amount of solar heat, water could not exist on their surfaces. He devotes a great deal of space to the evidence that there is no animal life on the moon, and taking this as proved, he uses it as a counter argument against the other side. They always urge that, the earth being inhabited, we must suppose the other planets to be so too; to which he replies:—We know that the moon is not inhabited though it has all the advantage of proximity to the sun that the earth has; why then should not other planets be equally uninhabited?

He then comes to Mars and admits that this planet is very like the earth so far as we can judge, and that it may therefore be inhabited, or as the author expresses it, 'may have been judged worthy of inhabitants by its Maker.' But he urges the small size of Mars, its coldness owing to distance from the sun, and that the annual melting of its polar ice-caps will keep it cold all through the summer. If there are animals they are probably of a low type like the saurians and iguanodons of our seas during the Wealden epoch; but, he argues, as even on our earth the long process of preparation for man was carried on for countless millions of years, we need not discuss whether there are intelligent beings on Mars till we have some better evidence that there are any living creatures at all.

Several of the early chapters are devoted to an attempt to minimise the difficulties of those religious persons who feel oppressed by the immensity and complexity of the material universe as revealed by modern astronomy; and by the almost infinite insignificance of man and his dwelling-place, the earth, in comparison with it, an insignificance vastly increased if not only the planets of the solar system, but also those which circle around the myriads of suns, are also theatres of life. And these persons are further disquieted because the very same facts are used by sceptics of various kinds in their attacks upon Christianity. Such writers point out the irrationality and absurdity of supposing that the Creator of all this unimaginable vastness of suns and systems, filling for all we know endless space, should take any *special* interest in so mean and pitiful a creature as man, the imperfectly developed inhabitant of one of the smaller worlds attached to a second or third-rate sun, a being whose whole history is one of war and bloodshed, of tyranny, torture, and death; whose awful record is pictured by himself in such books as Josephus' *History of the Jews*, the *Decline and Fall of the Roman Empire*, and even more forcibly summarised in that terrible picture of human fiendishness and misery, *The Martyrdom of Man*; while their character is indicated by one of the kindest and simplest of their poets in the restrained but expressive lines:—

'Man's inhumanity to man Makes countless thousands mourn.'

It is for such a being as this, they say, that God should have specially revealed His will some thousands of years ago, and finding that His commands were not obeyed, His will not fulfilled, yet ordained for their benefit the necessarily unique sacrifice of His Son, in order to save a small portion of these 'miserable sinners' from the natural and well-deserved consequence of their stupendous follies, their unimaginable crimes? Such a belief they maintain is too absurd, too incredible, to be held by any rational being, and it becomes even less credible and less rational if we maintain that there are countless other inhabited worlds.

It is very difficult for the religious man to make any

adequate reply to such an attack as this, and as a result many have felt their position to be untenable and have accordingly lost all faith in the special dogmas of orthodox Christianity. They feel themselves really to be between the horns of a dilemma. If there are myriads of other worlds, it seems incredible that they should each be the object of a special revelation and a special sacrifice. If, on the other hand, we are the only intelligent beings that exist in the material universe, and are really the highest creative product of a Being of infinite wisdom and power, they cannot but wonder at the vast apparent disproportion between the Creator and the created, and are sometimes driven to Atheism from the hopelessness of comprehending so mean and petty a result as the sole outcome of infinite power.

Whewell tells us that the great preacher, Dr. Chalmers, in his Astronomical Discourses, attempted a reply to these difficulties, but, in his opinion, not a very successful one; and a large part of his own work is devoted to the same purpose. His main point seems to be that we know too little of the universe to arrive at any definite conclusions on the question at issue, and that any ideas that we may have as to the purposes of the Creator in forming the vast system we see around us are almost sure to be erroneous. We must therefore be content to remain ignorant, and must rest satisfied in the belief that the Creator had a purpose although we are not yet permitted to know what it was. And to those who urge that in other worlds there may be other laws of nature which may render them quite as habitable by intelligent beings as our world is for us, he replies, that if we are to suppose new laws of nature in order to render each planet habitable, there is an end of all rational inquiry on the subject, and we may maintain and believe that animals may live on the moon without air or water, and on the sun exposed to heat which vaporises earths and metals.

His concluding argument, and perhaps one of his strongest, is that founded upon the dignity of man, as conferring a pre-eminence upon the planet which has produced him. 'If,' he says, 'man be not merely capable of Virtue and Duty, of universal Love and Self-Devotion, but be also immortal; if his being be of infinite duration, his soul created never to die; then, indeed, we may well say that one soul outweighs the whole unintelligent creation.' And then, addressing the religious world, he urges that, if, as they believe, God *has* redeemed man by the sacrifice of His Son, and *has* given to him a revelation of His will, then indeed no other conception is possible than that he is the sole and highest product of the universe. 'The elevation of millions of intellectual, moral, religious, spiritual creatures, to a destiny so prepared, consummated, and developed, is no unworthy occupation of all the capacities of space, time, and matter.' Then with a chapter on 'The Unity of the World,' and one on 'The Future,' neither of which contains anything which adds to the force of his argument, the book ends.

The publication of this able if rather vague and diffuse work, contesting popular opinions, was followed by a burst of indignant criticism on the part of a man of considerable eminence in some branches of physics—Sir David Brewster, but who was very inferior, both in general knowledge of science and in literary skill, to the writer whose views he opposed. The purport of the book in which he set forth his objections is indicated by its title—*More Worlds than One, the Creed of the Philosopher and the Hope of the Christian.* Though written with much force and conviction it appeals mainly to religious prejudices, and assumes throughout that every planet and star is a special creation, and that the peculiarities of each were designed for some special purpose. 'If,' he says, 'the moon had been destined to be merely a lamp to our earth, there was no occasion to variegate its surface with lofty mountains and extinct volcanoes, and cover it with large patches of matter that reflect different quantities of light and give its surface the appearance of continents and seas. It would have been a better lamp had it been a smooth piece of lime or of chalk.' It is, therefore, he thinks, prepared for inhabitants; and then he argues that all the other satellites are also inhabited. Again he says that 'when it was found that Venus was about the same size as the Earth, with mountains and valleys, days and nights, and years analogous to our own, the *absurdity* of believing that she had no inhabitants, when

31

no other rational purpose could be assigned for her creation, became an argument of a certain amount that she was, like the Earth, the seat of animal and vegetable life.' Then, when it was found that Jupiter was so gigantic 'as to require four moons to give him light, the argument from analogy that *he* was inhabited became stronger also, because it extended to *two* planets.' And thus each successive planet having certain points of analogy with the others becomes an additional argument; so that when we take account of all the planets, with atmosphere, and clouds, and arctic snows, and trade-winds, the argument from analogy becomes, he urges, very powerful;—'and the absurdity of the opposite opinion, that planets should have moons and no inhabitants, atmospheres with no creatures to breathe in them, and currents of air without life to be fanned, became a formidable argument which few minds, if any, could resist.'

The work is full of such weak and fallacious rhetoric and even, if possible, still weaker. Thus after describing double stars, he adds—'But no person can believe that two suns could be placed in the heavens for no other purpose than to revolve round their common centre of gravity'; and he concludes his chapter on the stars thus:—'Wherever there is matter there must be Life; Life Physical to enjoy its beauties—Life Moral to worship its Maker, and Life Intellectual to proclaim His wisdom and His power.' And again—'A house without tenants, a city without citizens, presents to our minds the same idea as a planet without life, and a universe without inhabitants. Why the house was built, why the city was founded, why the planet was made, and why the universe was created, it would be difficult even to conjecture.' Arguments of this kind, which in almost every case beg the question at issue, are repeated *ad nauseam*. But he also appeals to the Old Testament to support his views, by quoting the fine passage in the Psalms—'When I consider Thy heavens the work of Thy fingers, the moon and the stars which Thou hast ordained; what is man that Thou art mindful of him?' on which he remarks—'We cannot doubt that inspiration revealed to him [David] the magnitude, the distances, and the final cause, of the glorious spheres which fixed his admiration.' And after quoting various other passages from the prophets, all as he thinks supporting the same view, he sets forth the extraordinary idea as a confirmatory argument, that the planets or some of them are to be the future abode of man. For, he says—'Man in his future state of existence is to consist, as at present, of a spiritual nature residing in a corporeal frame. He must live, therefore, upon a material planet, subject to all the laws of matter.' And he concludes thus:—'If there is not room, then, on our globe for the millions of millions of beings who have lived and died on its surface, we can scarcely doubt that their future abode must be on some of the primary or secondary planets of the solar system, whose inhabitants have ceased to exist, or upon planets which have long been in a state of preparation, as our earth was, for the advent of intellectual life.'

It is pleasant to turn from such weak and trivial arguments to the only other modern works which deal at some length with this subject, the late Richard A. Proctor's *Other Worlds than Ours*, and a volume published five years later under the title—*Our Place Among Infinities*. Written as these were by one of the most accomplished astronomers of his day, remarkable alike for the acuteness of his reasoning and the clearness of his style, we are always interested and instructed even when we cannot agree with his conclusions. In the first work mentioned above, he assumes, like Sir David Brewster, the antecedent probability that the planets are inhabited and on much the same theological grounds. So strongly does he feel this that he continually speaks as if the planets *must* be inhabited unless we can show very good reason that they *cannot* be so, thus throwing the burden of proving a negative on his opponents, while he does not attempt to prove his positive contention that they are inhabited, except by purely hypothetical considerations as to the Creator's purpose in bringing them into existence.

But starting from this point he endeavours to show how Whewell's various difficulties may be overcome, and here he always appeals to astronomical or physical facts, and reasons well upon them. But he is quite honest; and, coming to the conclusion that Jupiter and Saturn, Uranus and Neptune, cannot be habitable, he adduces the evidence and plainly states the result. But then he thinks that the satellites of Jupiter and Saturn *may* be habitable, and if they may be, then he concludes that they *must*. One great oversight in his whole argument is, that he is satisfied with showing the possibility that life may exist now, but never deals with the question of whether life could have been developed from its earliest rudiments up to the production of the higher vertebrates and man; and this, as I shall show later, is the *crux* of the whole problem.

With regard to the other planets, after a careful examination of all that is known about them, he arrives at the conclusion that if Mercury is protected by a cloud-laden atmosphere of a peculiar kind it may possibly, but not probably, support high forms of animal life. But in the case of Venus and Mars he finds so much resemblance to and so many analogies with our earth, that he concludes that they almost certainly are so.

In the case of the fixed stars, now that we know by spectroscopic observations that they are true suns, many of which closely resemble our sun and give out light and heat as he does, Mr. Proctor argues, that 'The vast supplies of heat thus emitted by the stars not only suggest the conclusion that there must be worlds around these orbs for which these heat-supplies are intended, but point to the existence of the various forms of force into which heat may be transmuted.

We know that the sun's heat poured upon our earth is stored up in vegetable and animal forms of life; is present in all the phenomena of nature—in winds and clouds and rain, in thunder and lightning, storm and hail; and that even the works of man are performed by virtue of the solar heat-supplies. Thus the fact that the stars send forth heat to the worlds which circle around them suggests at once the thought that on those worlds there must exist animal and vegetable forms of life.' We may note that in the first part of this passage the presence of worlds or planets is 'suggested,' while later on 'the worlds which circle round them' is spoken of as if it were a proved fact from which the presence of vegetable and animal life may be inferred. A suggestion depending on a preceding suggestion is not a very firm basis for so vast and wide-reaching a conclusion.

In the second work referred to above there is one chapter entitled, 'A New Theory of Life in other Worlds,' where the author gives his more matured views of the question, which are briefly stated in the preface as being 'that the weight of evidence favours my theory of the (relative) paucity of worlds.' His views are largely founded on the theory of probabilities, of which subject he had made a special study. Taking first our earth, he shows that the period during which life has existed upon it is very small in comparison with that during which it must have been slowly forming and cooling, and its atmosphere condensing so as to form land and water on its surface. And if we consider the time the earth has been occupied by man, that is a very minute part, perhaps not the thousandth part, of the period during which it has existed as a planet. It follows that even if we consider only those planets whose physical condition seems to us to be such as to be able to sustain life, the chances are perhaps hundreds to one against their being at that particular stage when life has begun to be developed, or if it has begun has reached as high a development as on our earth.

With regard to the stars, the argument is still stronger, because the epochs required for their formation are altogether unknown, while as to the conditions required for the formation of planetary systems around them we are totally ignorant. To this I would add that we are equally ignorant as to the probability or even possibility of many of these suns producing planets which, by their position, size, atmosphere, or other physical conditions can possibly become life-producing worlds. And, as we shall see later, this point has been overlooked by all writers, including Mr. Proctor himself. His conclusion is, then, that although the worlds which possess life at all approaching that of our earth may be relatively few in number, yet considering the universe as practically infinite in extent, they may be really very numerous.

It has been necessary to give this sketch of the views of those who have written specially on the question of the Plurality of Worlds, because the works referred to have been very widely read and have influenced educated opinion throughout the world. Moreover, Mr. Proctor, in his last work on the subject, speaks of the theory as being 'identified with modern astronomy'; and in fact popular works still discuss it. But all these follow the same general line of argument as those already referred to, and the curious thing is that while overlooking many of the most essential conditions they often introduce others which are by no means essential—as, for instance, that the atmosphere must have the same proportion of oxygen as our own. They seem to think that if any of our quadrupeds or birds taken to another planet could not live there, no animals of equally high organisation could inhabit it; entirely overlooking the very obvious fact that, supposing, as is almost certain, that oxygen is necessary for life, then, whatever proportion of oxygen within certain limits was present, the forms of life that arose would necessarily be organised in adaptation to that proportion, which might be considerably less or greater than on the earth.

The present volume will show how extremely inadequate has been the treatment of this question, which involves a variety of important considerations hitherto altogether overlooked. These are extremely numerous and very varied in their character, and the fact that they all point to one conclusion—a conclusion which so far as I am aware no previous writer has reached—renders it at least worthy of the careful consideration of all unbiassed thinkers. The whole subject is one as to which no direct evidence is obtainable, but I venture to think that the convergence of so many probabilities and indications towards a single definite theory, intimately connected with the nature and destiny of man himself, raises this theory to a very much higher level of probability than the vague possibilities and theological suggestions which are the utmost that have been adduced by previous writers.

In order to make every step of my argument clearly intelligible to all educated readers, it will be necessary to refer continually to the marvellous extension of our knowledge of the universe obtained during the last half-century, and constituting what is termed the New Astronomy. The next chapter will therefore be devoted to a popular exposition of the new methods of research, so that the results reached, which will have to be referred to in succeeding chapters, may be not only accepted, but clearly understood.

CHAPTER III

THE NEW ASTRONOMY

During the latter half of the nineteenth century discoveries were made which extended the powers of astronomical research into entirely new and unexpected regions, comparable to those which were opened up by the discovery of the telescope more than two centuries before. The older astronomy for more than two thousand years was purely mechanical and mathematical, being limited to observation and measurement of the apparent motions of the heavenly bodies, and the attempts to deduce, from these apparent motions, their real motions, and thus determine the actual structure of the solar system. This was first done when Kepler established his three celebrated laws: and later, when Newton showed that these laws were necessary consequences of the one law of gravitation, and when succeeding observers and mathematicians proved that each fresh irregularity in the motions of the planets was explicable by a more thorough and minute application of the same laws, this branch of astronomy reached its highest point of efficiency and left very little more to be desired.

Then, as the telescope became successively improved, the

centre of interest was shifted to the surfaces of the planets and their satellites, which were watched and scrutinised with the greatest assiduity in order if possible to attain some amount of knowledge of their physical constitution and past history. A similar minute scrutiny was given to the stars and nebulæ, their distribution and grouping, and the whole heavens were mapped out, and elaborate catalogues constructed by enthusiastic astronomers in every part of the world. Others devoted themselves to the immensely difficult problem of determining the distances of the stars, and by the middle of the century a few such distances had been satisfactorily measured.

Thus, up to the middle of the nineteenth century it appeared likely that the future of astronomy would rest almost entirely on the improvement of the telescope, and of the various instruments of measurement by means of which more accurate determinations of distances might be obtained. Indeed, the author of the Positive Philosophy, Auguste Comte, felt so sure of this that he deprecated all further attention to the stars as pure waste of time that could never lead to any useful or interesting result. In his *Philosophical Treatise on Popular Astronomy* published in 1844, he wrote very strongly on this point. He there tells us that, as the stars are only accessible to us by sight they must always remain very imperfectly known. We can know little more than their mere existence. Even as regards so simple a phenomenon

41

as their temperature this must always be inappreciable to a purely visual examination. Our knowledge of the stars is for the most part purely negative, that is, we can determine only that they do *not* belong to our system. Outside that system there exists, in astronomy, only obscurity and confusion, for want of indispensable facts; and he concludes thus:—'It is, then, in vain that for half a century it has been endeavoured to distinguish two astronomies, the one solar the other sidereal. In the eyes of those for whom science consists of real laws and not of incoherent facts, the second exists only in name, and the first alone constitutes a true astronomy; and I am not afraid to assert that it will always be so.' And he adds that—'all efforts directed to this subject for half a century have only produced an accumulation of incoherent empirical facts which can only interest an irrational curiosity.'

Seldom has a confident assertion of finality in science received so crushing a reply as was given to the above statements of Comte by the discovery in 1860 (only three years after his death) of the method of spectrum-analysis which, in its application to the stars, has revolutionised astronomy, and has enabled us to obtain that very kind of knowledge which he declared must be for ever beyond our reach. Through it we have acquired accurate information as to the physics and chemistry of the stars and nebulæ, so that we now know really more of the nature, constitution, and temperature of the enormously distant suns which we distinguish by the general term stars, than we do of most of the planets of our own system. It has also enabled us to ascertain the existence of numerous invisible stars, and to determine their orbits, their rate of motion, and even, approximately, their mass. The despised stellar astronomy of the early part of the century has now taken rank as the most profoundly interesting department of that grand science, and the branch which offers the greatest promise of future discoveries. As the results obtained by means of this powerful instrument will often be referred to, a short account of its nature and of the principles on which it depends must here be given.

The solar spectrum is the band of coloured light seen in the rainbow and, partially, in the dew-drop, but more completely when a ray of sunlight passes through a prism—a piece of glass having a triangular section. The result is, that instead of a spot of white light we have a narrow band of brilliant colours which succeed each other in regular order, from violet at one end through blue, green, and yellow to red at the other. We thus see that light is not a simple and uniform radiation from the sun, but is made up of a large number of separate rays, each of which produces in our eyes the sensation of a distinct colour. Light is now explained as being due to vibrations of ether, that mysterious substance which not only permeates all matter, but which fills space at least as far as the remotest of the visible stars and nebulæ.

The exceedingly minute waves or vibrations of the ether produce all the phenomena of heat, light, and colour, as well as those chemical actions to which photography owes its wonderful powers. By ingenious experiments the size and rate of vibration of these waves have been measured, and it is found that they vary considerably, those forming the red light, which is least refracted, having a wave-length of about $1/_{326000}$ of an inch, while the violet rays at the other end of the spectrum are only about half that length or $1/_{630000}$ of an inch. The rate at which the vibrations succeed each other is from 302 millions of millions per second for the extreme red rays, to 737 millions of millions for those at the violet end of the spectrum. These figures are given to show the wonderful minuteness and rapidity of these heat and light waves on which the whole life of the world, and all our knowledge of other worlds and other suns, directly depends.

But the mere colours of the spectrum are not the most important part of it. Very early in the nineteenth century a close examination showed that it was everywhere crossed by black lines of various thicknesses, sometimes single, sometimes grouped together. Many observers studied them and made accurate drawings or maps showing their positions and thicknesses, and by combining several prisms, so that the beam of sunlight had to pass through them successively, a spectrum could be produced several feet long, and more than 3000 of these dark lines were counted in it. But what they were and how they were caused remained a mystery, till, in the year 1860, the German physicist Kirchhoff discovered the secret and gave to chemists and astronomers a new and quite unexpected engine of research.

It had already been observed that the chemical elements and various compounds, when heated to incandescence, produced spectra consisting of coloured lines or bands which were constant for each element, so that the elements could at once be recognised by their characteristic spectra; and it had also been noticed that some of these bands, especially the yellow band produced by sodium, corresponded in position with certain black lines in the solar spectrum. Kirchhoff's discovery consisted in showing that, when the light from an incandescent body passes through the same substance in a state of vapour or gas, so much of the light is absorbed that the coloured lines or bands become black. The mystery of more than half a century was thus solved; and the thousands of black lines in the solar spectrum were shown to be caused by the light from the incandescent matter of the sun's surface passing through the heated gases or vapours immediately above it, and thereby having the bright coloured lines of their spectra changed, by absorption, to comparative blackness.

Chemists and physicists immediately set to work examining the spectra of the elements, fixing the position of the several coloured lines or bands by accurate measurement, and comparing them with the dark lines of the solar spectrum.

The results were in the highest degree satisfactory. In a large proportion of the elements the coloured bands corresponded exactly with a group of dark lines in the spectrum of the sun, in which, therefore, the same terrestrial elements were proved to exist. Among the elements first detected in this manner were hydrogen, sodium, iron, copper, magnesium, zinc, calcium, and many others. Nearly forty of the elements have now been found in the sun, and it seems highly probable that all our elements really exist there, but as some are very rare and are present in very minute quantities they cannot be detected. Some of the dark lines in the sun were found not to correspond to any known element, and as this was thought to indicate an element peculiar to the sun it was named Helium; but quite recently it has been discovered in a rare mineral. Many of the elements are represented by a great number of lines, others by very few. Thus iron has more than 2000, while lead and potassium have only one each.

The value of the spectroscope both to the chemist in discovering new elements and to the astronomer in determining the constitution of the heavenly bodies, is so great, that it became of the highest importance to have the position of all the dark lines in the solar spectrum, as well as the bright lines of all the elements, determined with extreme accuracy, so as to be able to make exact comparisons between different spectra. At first this was done by means of very large-scale drawings showing the exact position of every dark or bright line. But this was found to be both inconvenient and not sufficiently exact; and it was therefore agreed to adopt the natural scale of the wave-lengths of the different parts of the spectrum, which by means of what are termed diffraction-gratings can now be measured with great accuracy. Diffraction-gratings are formed of a polished surface of hard metal ruled with excessively fine lines, sometimes as many as 20,000 to an inch. When sunlight falls upon one of these gratings it is reflected, and by interference of the rays from the spaces between the fine grooves, it is spread out into a beautiful and well-defined spectrum, which, when the lines are very close, is several yards in length. In these diffraction spectra many dark lines are seen which can be shown in no other way, and they also give a spectrum which is far more uniform than that produced by glass prisms in which minute differences in the composition of the glass cause some rays to be refracted more and others less than the normal amount.

The spectra produced by diffraction-gratings are double; that is, they are spread out on both sides of the central line of the ray which remains white, and the several coloured or dark lines are so clearly defined that they can be thrown on a screen at a considerable distance, giving a great length to the spectrum. The data for obtaining the wave-lengths are the distance apart of the lines, the distance of the screen, and the distance apart of the first pair of dark lines on each side of the central bright line. All these can be measured with extreme accuracy by means of telescopes with micrometers and other contrivances, and the result is an accuracy of determination of wave-lengths which can probably not be equalled in any other kind of measurement.

As the wave-lengths are so excessively minute, it has been found convenient to fix upon a still smaller unit of measurement, and as the millimetre is the smallest unit of the metric system, the ten-millionth of a millimetre (technically termed 'tenth meter') is the unit adopted for the measurement of wave-lengths, which is equal to about the 250 millionth of an inch. Thus the wave-lengths of the red and blue lines characteristic of hydrogen are 6563.07 and 4861.51 respectively. This excessively minute scale of wave-lengths, once determined by the most refined measurement, is of very great importance. Having the wave-lengths of any two lines of a spectrum so determined, the space between them can be laid down on a diagram of any length, and all the lines that occur in any other spectrum between these two lines can be marked in their exact relative positions. Now, as the visible spectrum consists of about 300,000 rays of light, each of different wave-lengths and therefore of different refrangibilities, if it is laid down on such a scale as to be of a length of 3000 inches (250 feet), each wave-length will be $^1/_{100}$ of an inch long, a space easily visible by the naked eye.

The possession of an instrument of such wonderful delicacy, and with powers which enable it to penetrate into the inner constitution of the remotest orbs of space, rendered it possible, within the next quarter of a century, to establish what is practically a new science—Astrophysics—often popularly termed the New Astronomy. A brief outline of the main achievements of this science must now be given.

The first great discovery made by Spectrum analysis, after the interpretation of the sun's spectrum had been obtained, was, the real nature of the fixed stars. It is true they had long been held by astronomers to be suns, but this was only an opinion of the accuracy of which it did not seem possible to obtain any proof. The opinion was founded on two facts—their enormous distance from us, so great that the whole diameter of the earth's orbit did not lead to any apparent change of their relative positions, and their intense brilliancy which at such distances could only be due to an actual size and splendour comparable with our sun. The spectroscope at once proved the correctness of this opinion. As one after another was examined, they were found to exhibit spectra of the same general type as that of the sun—a band of colours crossed by dark lines. The very first stars examined by Sir William Huggins showed the existence of nine or ten of our elements. Very soon all the chief stars of the heavens were spectroscopically examined, and it was found that they could be classed in three or four groups. The first and largest group contains more than half the visible stars, and a still larger proportion of the most brilliant, such as Sirius, Vega, Regulus, and Alpha Crucis in the Southern Hemisphere. They are characterised by a white or bluish light, rich in the ultra-violet rays, and their spectra are distinguished by the breadth and intensity of the four dark bands due to the absorption of hydrogen, while the various black lines which indicate metallic vapours are comparatively few, though hundreds of them can be discovered by careful examination.

The next group, to which Capella and Arcturus belong, is also very numerous, and forms the solar type of stars. Their light is of a yellowish colour, and their spectra are crossed throughout by innumerable fine dark lines more or less closely corresponding with those in the solar spectrum.

The third group consists of red and variable stars, which are characterised by fluted spectra. Such spectra show like a range of Doric columns seen in perspective, the red side being that most illuminated.

The last group, consisting of few and comparatively small stars, has also fluted spectra, but the light appears to come from the opposite direction.

These groups were established by Father Secchi, the Roman astronomer, in 1867, and have been adopted with some modifications by Vogel of the Astrophysical Observatory at Potsdam. The exact interpretation of these different spectra is somewhat uncertain, but there can be little doubt that they coincide primarily with differences of temperature and with corresponding differences in the composition and extent of the absorptive atmospheres. Stars with fluted spectra indicate the presence of vapours of the metalloids or of compound substances, while the reversed flutings indicate the presence of carbon. These conclusions have been reached by careful laboratory experiments which are now carried on at the same time as the spectral examination of the stars and other heavenly bodies, so that each peculiarity of their spectra, however puzzling and apparently unmeaning, has been usually explained, by being shown to indicate certain conditions of chemical constitution or of temperature.

But whatever difficulty there may be in explaining details, there remains no doubt whatever of the fundamental fact that all the stars are true suns, differing no doubt in size, and their stage of development as indicated by the colour or intensity of their light or heat, but all alike possessing a photosphere or light-emitting surface, and absorptive atmospheres of various qualities and density.

Innumerable other details, such as the often contrasted colours of double stars, the occasional variability of their spectra, their relations to the nebulæ, the various stages of their development and other problems of equal interest, have occupied the continued attention of astronomers, spectroscopists, and chemists; but further reference to these difficult questions would be out of place here. The present sketch of the nature of spectrum-analysis applied to the stars is for the purpose of making its principle and method of observation intelligible to every educated reader, and to illustrate the marvellous precision and accuracy of the results attained by it. So confident are astronomers of this accuracy that nothing less than *perfect correspondence* of the various bright lines in the spectrum of an element in the laboratory with the dark lines in the spectrum of the sun or of a star is required before the presence of that element is accepted as proved. As Miss Clerke tersely puts it—'Spectroscopic coincidences admit of no compromise. Either they are absolute or they are worthless.'

Measurement of Motion in the Line of Sight

We must now describe another and quite distinct application of the spectroscope, which is even more marvellous than that already described. It is the method of measuring the rate of motion of any of the visible heavenly bodies in a direction either directly towards us, or directly away from us, technically described as 'radial motion,' or by the expression—'in the line of sight.' And the extraordinary thing is that this power of measurement is altogether independent of distance, so that the rate of motion in miles per second of the remotest of the fixed stars, if sufficiently bright to show a distinct spectrum, can be measured with as much certainty and accuracy as in the case of a much nearer star or a planet.

In order to understand how this is possible we have again to refer to the wave-theory of light; and the analogy of other wave-motions will enable us better to grasp the principle on which these calculations depend. If on a nearly calm day we count the waves that pass each minute by an anchored steamboat, and then travel in the direction the waves come from, we shall find that a larger number pass us in the same time. Again, if we are standing near a railway, and an engine comes towards us whistling, we shall notice that it changes its tone as it passes us; and as it recedes the sound will be in a lower key, although the engine may be at exactly the same distance from us as when it was approaching. Yet the sound does not change to the ear of the engine driver, the cause of the change being that the sound-waves reach us in quicker succession as the source of the waves is approaching us than when it is retreating from us. Now, just as the pitch of a note depends upon the rapidity with which the successive air-vibrations reach our ear, so does the colour of a particular part of the spectrum depend upon the rapidity with which the ethereal waves which produce colour reach our eyes; and as this rapidity is greater when the source of the light is approaching than when it is receding from us, a slight shifting of the position of the coloured bands, and therefore of the dark lines, will occur, as compared with their position in the spectrum of the sun or of any stationary source of light, if there is any motion sufficient in amount to produce a perceptible shift.

That such a change of colour would occur was pointed out by Professor Doppler of Prague in 1842, and it is hence usually spoken of as the 'Doppler principle'; but as the changes of colour were so minute as to be impossible of measurement it was not at that time of any practical importance in astronomy. But when the dark lines in the spectrum were carefully mapped, and their positions determined with minute accuracy, it was seen that a means of measuring the changes produced by motion in the line of sight existed, since the position of any of the dark or coloured lines in the spectra of the heavenly bodies could be compared with those of the corresponding lines produced artificially in the laboratory. This was first done in 1868 by Sir William Huggins, who, by the use of a very powerful spectroscope constructed for the purpose, found that such a change did occur in the case of many stars, and that their rate of motion towards us or away from us—the radial motion—could be calculated. As the actual distance of some of these stars had been measured, and their change of position annually (their proper motion) determined, the additional factor of the amount of motion in the direction of our line of sight completed the data required to fix their true line of motion among the other stars. The accuracy of this method under favourable conditions and with the best instruments is very great, as has been proved by those cases in which we have independent means of calculating the real motion. The motion of Venus towards or away from us can be calculated with great accuracy for any period, being a resultant of the combined motions of the planet and of our earth in their respective orbits. The radial motions of Venus were determined at the Lick Observatory in August and September 1890, by spectroscopic observations, and also by calculation, to be as follows:—

		By Observation.	By Calculation.
Aug.	16th.	7.3 miles per second.	8.1 miles per second.
"	22nd.	8.9 " " "	8.2 " " "
"	30th.	7.3 " " "	8.3 " " "
Sep.	3rd.	8.3 " " "	8.3 " " "
"	4th.	8.2 " " "	8.3 " " "

showing that the maximum error was only one mile per second, while the mean error was about a quarter of a mile. In the case of the stars the accuracy of the method has been tested by observations of the same star at times when the earth's motion in its orbit is towards or away from the star, whose apparent radial velocity is, therefore, increased or diminished by a known amount. Observations of this kind were made by Dr. Vogel, Director of the Astrophysical Observatory at Potsdam, showing, in the case of three stars, of which ten observations were taken, a mean error of about two miles per second; but as the stellar motions are more rapid than those of the planets, the proportionate error is no greater than in the example given above.

The great importance of this mode of determining the real motion of the stars is, that it gives us a knowledge of the scale on which such motions are progressing; and when in the course of time we discover whether any of their paths are rectilinear or curved, we shall be in a position to learn something of the nature of the changes that are going on and of the laws on which they depend.

Invisible Stars and Imperceptible Motions

But there is another result of this power of determining radial motion which is even more unexpected and marvellous, and which has extended our knowledge of the stars in quite a new direction. By its means it is possible to determine the existence of invisible stars and to measure the rate of otherwise imperceptible motions; that is of stars which are invisible in the most powerful modern telescopes, and whose motions have such a limited range that no telescope can detect them.

Double or binary stars forming systems which revolve around their common centre of gravity were discovered by Sir William Herschel, and very great numbers are known; but in most cases their periods of revolution are long, the shortest being about twelve years, while many extend to several hundred years. These are, of course, all visible binaries, but many are now known of which one star only is visible while the other is either non-luminous or is so close to its companion that they appear as a single star in the most powerful telescopes. Many of the variable stars belong to the former class, a good example of which is Algol in the constellation Perseus, which changes from the second to the fourth magnitude in about four and a half hours, and in about four and a half hours more regains its brilliancy till its next period of obscuration which occurs regularly every two days and twenty-one hours. The name Algol is from the Arabic *Al Ghoul*, the familiar 'ghoul' of the Arabian Nights, so named—'The Demon'—from its strange and weird behaviour.

It had long been conjectured that this obscuration was due to a dark companion which partially eclipsed the bright star at every revolution, showing that the plane of the orbit of the pair was almost exactly directed towards us. The application of the spectroscope made this conjecture a certainty. At an equal time before and after the obscuration, motion in the line of sight was shown, towards and away from us, at a rate of twenty-six miles per second. From these scanty data and the laws of gravitation which fix the period of revolution of planets at various distances from their centres of revolution, Professor Pickering of the Harvard Observatory was able to arrive at the following figures as highly probable, and they may be considered to be certainly not far from the truth.

Diameter of Algol,	1,061,000	miles.
Diameter of dark companion,	830,000	"
Distance between their centres,	3,230,000	"
Orbital speed of Algol,	26.3	miles per sec.
Orbital speed of companion,	55.4	" " "
Mass of Algol,	$4/9$	mass of our S
Mass of companion,	$2/9$	" " "

When it is considered that these figures relate to a pair of stars only one of which has ever been seen, that the orbital motion even of the visible star cannot be detected in the most powerful telescopes, when, further, we take into account the enormous distance of these objects from us, the great results of spectroscopic observation will be better appreciated.

But besides the marvel of such a discovery by such simple means, the facts discovered are themselves in the highest degree marvellous. All that we had known of the stars through telescopic observation indicated that they were at very great distances from each other however thickly they may appear scattered over the sky. This is the case even with close telescopic double stars, owing to their enormous remoteness from us. It is now estimated that even stars of the first magnitude are, on a general average, about eighty millions of millions of miles distant; while the closest double stars that can be distinctly separated by large telescopes are about half a second apart. These, if at the above distance, will be about 1500 millions of miles from each other. But in the case of Algol and its companion, we have two bodies both larger than our sun, yet with a distance of only $2^1/_4$ millions of miles between their surfaces, a distance not much exceeding their combined diameters. We should not have anticipated that such huge bodies could revolve so closely to each other, and as we now know that the neighbourhood of our sun—and probably of all suns—is full of meteoric and cometic matter, it would seem probable that in the case of two suns so near together the quantity of such matter would be very great, and would lead probably by continued collisions to increase of their bulk, and perhaps to their final coalescence into a single giant orb. It is said that a Persian astronomer in the tenth century calls Algol a red star, while it is now white or somewhat yellowish. This would imply an increase of temperature caused by collisions or friction, and increasing proximity of the pair of stars.

A considerable number of double stars with dark companions have been discovered by means of the spectroscope, although their motion is not directly in the line of sight, and therefore there is no obscuration. In order to discover such pairs the spectra of large numbers of stars are taken on photographic plates every night and for considerable periods—for a year or for several years. These plates are then carefully examined with a high magnifying power to discover any periodical displacement of the lines, and it is astonishing in how large a number of cases this has been found to exist and the period of revolution of the pair determined.

But besides discovering double stars of which one is dark and one bright, many pairs of bright stars have been discovered by the same means. The method in this case is rather different. Each component star, being luminous, will give a separate spectrum, and the best spectroscopes are so powerful that they will separate these spectra when the stars are at their maximum distance although no telescope in existence, or ever likely to be made, can separate the component stars. The separation of the spectra is usually shown by the most prominent lines becoming double and then after a time single, indicating that the plane of revolution is more or less obliquely towards us, so that the two stars if visible would appear to open out and then get nearer together every revolution. Then, as each star alternately approaches and recedes from us the radial velocity of each can be determined, and this gives the relative mass. In this way not only doubles, but triple and multiple systems, have been discovered. The stars proved to be double by these two methods are so numerous that it has been estimated by one of the best observers that about one star in every thirteen shows inequality in its radial motion and is therefore really a double star.

The Nebulæ

One other great result of spectrum-analysis, and in some respects perhaps the greatest, is its demonstration of the fact that true nebulæ exist, and that they are not all star-clusters so remote as to be irresolvable, as was once supposed. They are shown to have gaseous spectra, or sometimes gaseous and stellar spectra combined, and this, in connection with the fact that nebulæ are frequently aggregated around nebulous stars or groups of stars, renders it certain that the nebulæ are in no way separated in space from the stars, but that they constitute essential parts of one vast stellar universe. There is, indeed, good reason to believe that they are really the material out of which stars are made, and that in their forms, aggregations, and condensations, we can trace the very process of evolution of stars and suns.

Photographic Astronomy

But there is yet another powerful engine of research which the new astronomy possesses, and which, either alone or in combination with the spectroscope, had produced and will yet produce in the future an amount of knowledge of the stellar universe which could never be attained by any other means. It has already been stated how the discovery of new variable and binary stars has been rendered possible by the preservation of the photographic plates on which the spectra are self-recorded, night after night, with every line, whether dark or coloured, in true position, so as to bear magnification, and, by comparison with others of the series, enabling the most minute changes to be detected and their amount accurately measured. Without the preservation of such comparable records, which is in no other way possible, by far the larger portion of spectroscopic discoveries could never have been made.

But there are two other uses of photography of quite a different nature which are equally and perhaps in their final outcome may be far more important. The first is, that by the use of the photographic plate the exact positions of scores, hundreds, or even thousands of stars can be self-mapped simultaneously with extreme accuracy, while any number of copies can be made of these star-maps. This entirely obviates the necessity for the old method of fixing the position of each star by repeated measurement by means of very elaborate instruments, and their registration in laborious and expensive catalogues. So important is this now seen to be, that specially constructed cameras are made for stellar photography, and by means of the best kinds of equatorial mounting are made to revolve slowly so that the image of each star remains stationary upon the plate for several hours.

Arrangements have been now made among all the chief observatories of the world to carry out a photographic survey of the heavens with identical instruments, so as to produce maps of the whole star-system on the same scale. These will serve as fixed data for future astronomers, who will thus be able to determine the movements of stars of all magnitudes with a certainty and accuracy hitherto unattainable.

The other important use of photography depends upon the fact that with a longer exposure within certain limits we increase the light-collecting power. It will surprise many persons to learn that an ordinary good portrait-camera with a lens three or four inches in diameter, if properly mounted so that an exposure of several hours can be made, will show stars so minute that they are invisible even in the great Lick telescope. In this way the camera will often reveal double-stars or small groups which can be made visible in no other way.

Such photographs of the stars are now constantly reproduced in works on Astronomy and in popular magazine articles, and although some of them are very striking, many persons are disappointed with them, and cannot understand their great value, because each star is represented by a white circle often of considerable size and with a somewhat undefined outline, not by a minute point of light as stars appear in a good telescope. But the essential matter in all such photographs is not so much the smallness, as the roundness, of the star-images, as this proves the extreme precision with which the image of every star has been kept by the clockwork motion of the instrument on the same point of the plate during the whole exposure. For example, in the fine photograph of the Great Nebula in Andromeda, taken 29th December 1888, by Dr. Isaac Roberts, with an exposure of four hours, there are probably over a thousand stars large and small to be seen, every one represented by an almost exactly circular white dot of a size dependent on the magnitude of the star. These round dots can be bisected by the cross hairs of a micrometer with very great accuracy, and thus the distance between the centres of any of the pairs, as well as the direction of the line joining their centres, can be determined as accurately as if each was represented by a point only. But as a minute white speck would be almost invisible on the maps, and would convey no information as to the approximate magnitude of the star, mistakes would be much more easily made, and it would probably be found necessary to surround each star with a circle to indicate its magnitude, and to enable it to be easily seen. It is probable, therefore, that the supposed defect is really an important advantage. The above-mentioned photograph is beautifully reproduced in Proctor's *Old and New Astronomy*, published after his greatly lamented death.

But besides the amount of altogether new knowledge obtained by the methods of research here briefly explained, a great deal of light has been thrown on the distribution of the stars as a whole, and hence on the nature and extent of the stellar universe, by a careful study of the materials obtained by the old methods, and by the application of the doctrine of probabilities to the observed facts. In this way alone some very striking results have been reached, and these have been supported and strengthened by the newer methods, and also by the use of new instruments in the measurement of stellar distances. Some of these results bear so closely and directly upon the special subject of the present volume, that our next chapter must be devoted to a consideration of them.

CHAPTER IV

THE DISTRIBUTION OF THE STARS

If we look at the heavens on a clear, moonless night in winter, and from a position embracing the entire horizon, the scene is an inexpressibly grand one. The intense sparkling brilliancy of Sirius, Capella, Vega, and other stars of the first magnitude; their striking arrangement in constellations or groups, of which Orion, the Great Bear, Cassiopeiæ, and the Pleiades, are familiar examples; and the filling up between these by less and less brilliant points down to the limit of vision, so as to cover the whole sky with a scintillating tracery of minute points of light, convey together an idea of such confused scattering and such enormous numbers, that it seems impossible to count them or to reduce them to systematic order. Yet this was done for all except the faintest stars by Hipparchus, 134 B.C., who catalogued and fixed the positions of more than 1000 stars, and this is about the number, down to the fifth magnitude, visible in the latitude of Greece. A recent enumeration of all the stars visible to the naked eye, under the most favourable conditions and by the best eyesight, has been made by the American astronomer, Pickering. His numbers are—for the Northern

Hemisphere 2509, and for the Southern Hemisphere 2824, thus showing a somewhat greater richness in the southern celestial hemisphere. But as this difference is due entirely to a preponderance of stars between magnitudes $5^1/_2$ and 6, that is, just on the limits of vision, while those down to magnitude $5^1/_2$ are more numerous by 85 in the Northern Hemisphere, Professor Newcomb is of opinion that there is no real superiority of numbers of visible stars in one hemisphere over the other. Again, the total number of the visible stars by the above enumeration is 5333. But this includes stars down to 6.2 magnitude, while it is generally considered that magnitude 6 marks the limit of visibility. On a re-examination of all the materials, the Italian astronomer Schiaparelli concludes that the total number of stars down to the sixth magnitude is 4303; and they seem to be about equally divided between the northern and southern skies.

THE MILKY WAY

But besides the stars themselves, a most conspicuous object both in the northern and southern hemisphere is that wonderful irregular belt of faintly diffused light termed the Milky Way or Galaxy. This forms a magnificent arch across the sky, best seen in the autumn months in our latitude. This arch, while following the general course of a great circle round the heavens, is extremely irregular in detail, sometimes being single, sometimes double, sending off occasional branches or offshoots, and also containing in its very midst dark rifts, spots, or patches, where the black background of almost starless sky can be seen through it. When examined through an opera-glass or small telescope quantities of stars are seen on the luminous background, and with every increase in the size and power of the telescope more and more stars become visible, till with the largest and best modern instruments the whole of the Galaxy seems densely packed with them, though still full of irregularities, wavy streams of stars, and dark rifts and patches, but always showing a faint nebulous background as if there remained other myriads of stars which a still higher optical power would reveal.

The relations of this great belt of telescopic stars to the rest of the star-system have long interested astronomers, and many have attempted its solution. By a system of gauging, that is counting all the stars that passed over the field of his telescope in a certain time, Sir William Herschel was the first who made a systematic effort to determine the shape of the stellar universe. From the fact that the number of stars increased rapidly as the Milky Way was approached from whatever direction, while in the Galaxy itself the numbers visible were at once more than doubled, he formed the idea that the shape of the entire system must be that of a highly compressed very broad mass or ring rather less dense towards the centre where our sun was situated. Roughly speaking, the form was likened to a flat disc or grindstone, but of irregular thickness, and split in two on one side where it appears to be double. The immense quantity of the stars which formed it was supposed to be due to the fact that we looked at it edgewise through an immense depth of stars; while at right angles to its direction when looking towards what is termed the pole of the Galaxy, and also in a less degree when looking obliquely, we see out into space through a much thinner stratum of stars, which thus seem on the average to be very much farther apart.

But, in the latter part of his life, Sir William Herschel realised that this was not the true explanation of the features presented by the Galaxy. The brilliant spots and patches in it, the dark rifts and openings, the narrow streams of light often bounded by equally narrow streams or rifts of darkness, render it quite impossible to conceive that this complex luminous ring has the form of a compressed disc extending in the direction in which we see it to a distance many times greater than its thickness. In one very luminous cluster Herschel thought that his telescope had penetrated to regions twenty times as far off as the more brilliant stars forming the nearer portions of the same object. Now, in the case of the Magellanic clouds, which are two roundish nebular patches of large size some distance from the Milky Way in the Southern Hemisphere and looking like detached portions of it, Sir John Herschel himself has shown that any such interpretation of its form is impossible; because it requires us to suppose that in both these cases we see, not rounded masses of a roughly globular shape, but immensely long cones or cylinders, placed in such a direction that we see only the ends of them. He remarks that one such object so situated would be an extraordinary coincidence, but that there should be two or many such is altogether out of the question. But in the Milky Way there are hundreds or even thousands of such spots or masses of exceptional brilliancy or exceptional darkness; and, if the form of the Galaxy is that of a disc many times broader than thick, and which we see edgewise, then every one of these patches and clusters, and all the narrow winding streams of bright light or intense blackness, must be really excessively long cylinders, or tunnels, or deep curving laminæ, or narrow fissures. And every one of these, which are to be found in every part of this vast circle of luminosity, must be so arranged as to be exactly turned towards our sun. The weight of this argument, which has been most forcibly and clearly set forth by the late Mr. R.A. Proctor, in his very instructive volume *Our Place among Infinities*, is now generally admitted by astronomers, and the natural conclusion is that the form of the Milky Way is that of a vast irregular ring, of which the section at any part is, roughly speaking, circular; while the many narrow rifts or lanes or openings where we seem to be able to see completely through it to the darkness of outer space beyond, render it probable that in those directions its thickness is less instead of greater than its apparent width, that is, that we see the broader side rather than the narrow edge of it.

Before entering on the consideration of the relations which the bulk of the stars we see scattered over the entire vault of heaven bear to this great belt of telescopic stars, it will be advisable to give a somewhat full description of the Galaxy itself, both because it is not often delineated on star-maps with sufficient accuracy, or so as to show its wonderful intricacies of structure, and also because it constitutes the fundamental phenomenon upon which the argument set forth in this volume primarily rests. For this purpose I shall use the description of it given by Sir John Herschel in his *Outlines of Astronomy*, both because he, of all the astronomers of the last century, had studied it most thoroughly, in the northern and in the southern hemispheres, by eye-observation and with the aid of telescopes of great power and admirable quality; and also because, amid the throng of modern works and the exciting novelties of the last thirty years, his instructive volume is, comparatively speaking, very little known. This precise and careful description will also be of service to any of my readers who may wish to form a closer personal acquaintance with this magnificent and intensely interesting object, by examining its peculiarities of form and beauties of structure either with the naked eye, or with the aid of a good opera-glass, or with a small telescope of good defining power.

A Description of the Milky Way

Sir John Herschel's description is as follows:—'The course of the Milky Way as traced through the heavens by the unaided eye, neglecting occasional deviations and following the line of its greatest brightness as well as its varying breadth and intensity will permit, conforms, as nearly as the indefiniteness of its boundary will allow it to be fixed, to that of a great circle inclined at an angle of about 63° to the equinoctial, and cutting that circle in Right Ascension 6h. 47m. and 18h. 47m., so that its northern and southern poles respectively are situated in Right Ascension 12h. 47m., North Polar Distance 63°, and R.A. 0h. 47m., NPD. 117°. Throughout the region where it is so remarkably subdivided, this great circle holds an intermediate situation between the two great streams; with a nearer approximation however to the brighter and continuous stream than to the fainter and interrupted one. If we trace its course in order of right ascension, we find it traversing the constellation Cassiopeiæ, its brightest part passing about two degrees to the north of the star Delta of that constellation. Passing thence between Gamma and Epsilon Cassiopeiæ, it sends off a branch to the south-preceding side, towards Alpha Persei, very conspicuous as far as that star, prolonged faintly towards Eta of the same constellation, and possibly traceable towards the Hyades and Pleiades as remote outliers. The main stream, however (which is here very faint), passes on through Auriga, over the three remarkable stars, Epsilon, Zeta, Eta, of that constellation called the Hædi, preceding Capella, between the feet of Gemini and the horns of the Bull (where it intersects the ecliptic nearly in the Solstitial Colure) and thence over the club of Orion to the neck of Monoceros, intersecting the equinoctial in R.A. 6h. 54m. Up to this point, from the offset in Perseus, its light is feeble and indefinite, but thenceforward it receives a gradual accession of brightness, and where it passes through the shoulder of Monoceros and over the head of Canis Major it presents a broad, moderately bright, very uniform, and to the naked eye, starless stream up to the point where it enters the prow of the ship Argo, nearly on the southern tropic. Here it again subdivides (about the star *m* Puppis), sending off a narrow and winding branch on the preceding side as far as Gamma Argûs, where it terminates abruptly. The main stream pursues its southward course to the 123rd parallel of NPD., where it diffuses itself broadly and again subdivides, opening out into a wide fan-like expanse, nearly 20° in breadth, formed of interlacing branches, which all terminate abruptly, in a line drawn nearly through Lambda and Gamma Argûs.

73

'At this place the continuity of the Milky Way is interrupted by a wide gap, and where it recommences on the opposite side it is by a somewhat similar fan-shaped assemblage of branches which converge upon the bright star Eta Argûs. Thence it crosses the hind feet of the Centaur, forming a curious and sharply-defined semicircular concavity of small radius, and enters the Cross by a very bright neck or isthmus of not more than three or four degrees in breadth, being the narrowest portion of the Milky Way. After this it immediately expands into a broad and bright mass, enclosing the stars Alpha and Beta Crucis and Beta Centauri, and extending almost up to Alpha of the latter constellation. In the midst of this bright mass, surrounded by it on all sides, and occupying about half its breadth, occurs a singular dark pear-shaped vacancy, so conspicuous and remarkable as to attract the notice of the most superficial gazer and to have acquired among the early southern navigators the uncouth but expressive appellation of the *coal-sack*. In this vacancy, which is about 8° in length and 5° broad, only one very small star visible to the naked eye occurs, though it is far from devoid of telescopic stars, so that its striking blackness is simply due to the effect of contrast with the brilliant ground with which it is on all sides surrounded. This is the place of nearest approach of the Milky Way to the South Pole. Throughout all this region its brightness is very striking, and when compared with that of its more northern course already

74

traced, conveys strongly the impression of greater proximity, and would almost lead to a belief that our situation as spectators is separated on all sides by a considerable interval from the dense body of stars composing the Galaxy, which in this view of the subject would come to be considered as a flat ring or some other re-entering form of immense and irregular breadth and thickness, within which we are excentrically situated, nearer to the southern than to the northern part of its circuit.

'At Alpha Centauri the Milky Way again subdivides, sending off a great branch of nearly half its breadth, but which thins off rapidly, at an angle of about 20° with its general direction to Eta and *d* Lupi, beyond which it loses itself in a narrow and faint streamlet. The main stream passes on increasing in breadth to Gamma Normæ, where it makes an abrupt elbow and again subdivides into one principal and continuous stream of very irregular breadth and brightness, and a complicated system of interlaced streaks and masses, which covers the tail of Scorpio, and terminates in a vast and faint effusion over the whole extensive region occupied by the preceding leg of Ophiuchus, extending northward to the parallel of 103° NPD., beyond which it cannot be traced; a wide interval of 14°, free from all appearance of nebulous light, separating it from the great branch on the north side of the equinoctial of which it is usually represented as a continuation.

'Returning to the point of separation of this great branch from the main stream, let us now pursue the course of the latter. Making an abrupt bend to the following side, it passes over the stars Iota Aræ, Theta and Iota Scorpii, and Gamma Tubi to Gamma Sagittarii, where it suddenly collects into a vivid oval mass about 6° in length and 4° in breadth, so excessively rich in stars that a very moderate calculation makes their number exceed 100,000. Northward of this mass, this stream crosses the ecliptic in longitude about 276°, and proceeding along the bow of Sagittarius into Antinous has its course rippled by three deep concavities, separated from each other by remarkable protuberances, of which the larger and brighter forms the most conspicuous patch in the southern portion of the Milky Way visible in our latitudes.

'Crossing the equinoctial at the 19th hour of R.A., it next runs in an irregular, patchy, and winding stream through Aquila, Sagitta, and Vulpecula up to Cygnus; at Epsilon of which constellation its continuity is interrupted, and a very confused and irregular region commences, marked by a broad dark vacuity, not unlike the southern "coal-sack," occupying the space between Epsilon, Alpha, and Gamma Cygni, which serves as a kind of centre for the divergence of three great streams; one, which we have already traced; a second, the continuation of the first (across the interval) from Alpha northward, between Lacerta and the head of Cepheus to the point in Cassiopeiæ whence we set out, and

76

a third branching off from Gamma Cygni, very vivid and conspicuous, running off in a southern direction through Beta Cygni, and *s* Aquilæ almost to the equinoctial, where it loses itself in a region thinly sprinkled with stars, where in some maps the modern constellation Taurus Poniatowski is placed. This is the branch which, if continued across the equinoctial, might be supposed to unite with the great southern effusion in Ophiuchus already noticed. A considerable offset, or protuberant appendage, is also thrown off by the northern stream from the head of Cepheus directly towards the pole, occupying the greater part of the quartile formed by Alpha, Beta, Iota, and Delta of that constellation.'

To complete this careful, detailed description of the Milky Way, it will be well to add a few passages from the same work as to its telescopic appearance and structure.

'When examined with powerful telescopes, the constitution of this wonderful zone is found to be no less various than its aspect to the naked eye is irregular. In some regions the stars of which it is composed are scattered with remarkable uniformity over immense tracts, while in others the irregularity of their distribution is quite as striking, exhibiting a rapid succession of closely clustering rich patches separated by comparatively poor intervals, and indeed in some instances by spaces absolutely dark *and completely void of any star*, even of the smallest telescopic magnitude. In some places not more than 40 or 50 stars on an average occur in a gauge-field of 15', while in others a similar average gives a result of 400 or 500. Nor is less variety observable in the character of its different regions in respect of the magnitudes of the stars they exhibit, and the proportional numbers of the larger and smaller magnitudes associated together, than in respect of their aggregate numbers. In some, for instance, extremely minute stars occur in numbers so moderate as to lead us irresistibly to the conclusion that in these regions we see *fairly through* the starry stratum, since it is impossible otherwise that the numbers of the smaller magnitudes should not go on continually increasing ad infinitum. In such cases, moreover, the ground of the heavens is for the most part perfectly dark, which again would not be the case if innumerable multitudes of stars, too minute to be individually discernible, existed beyond. In other regions we are presented with the phænomenon of an almost uniform degree of brightness of the individual stars, accompanied with a very even distribution of them over the ground of the heavens, both the larger and smaller magnitudes being strikingly deficient. In such cases it is equally impossible not to perceive that we are looking *through* a sheet of stars nearly of a size, and of no great thickness compared with the distance which separates them from us. Were it otherwise we should be driven to suppose the more distant stars uniformly the larger, so as to compensate by their greater intrinsic brightness for their greater distance, a supposition contrary to all probability....

'Throughout by far the larger portion of the extent of the Milky Way in both hemispheres, the general blackness of the ground of the heavens on which its stars are projected, and the absence of that innumerable multitude and excessive crowding of the smallest visible magnitudes, and of glare produced by the aggregate light of multitudes too small to affect the eye singly, must, we think, be considered unequivocal indications that its dimensions in *directions where these conditions obtain* are not only not infinite, but that the space-penetrating power of our telescopes suffices fairly to pierce through and beyond it.'

In the above-quoted passage the italics are those of Sir John Herschel himself, and we see that he drew the very same conclusions from the facts he describes, and for much the same reasons, as Mr. Proctor has drawn from the observations of Sir William Herschel; and, as we shall see, the best astronomers to-day have arrived at a similar result, from the additional facts at their disposal, and in some cases from fresh lines of argument.

The Stars in Relation to the Milky Way

Sir John Herschel was so impressed with the form, structure, and immensity of the Galactic Circle, as he sometimes terms it, that he says (in a footnote p. 575, 10th ed.), 'This circle is to sidereal what the invariable ecliptic is to planetary astronomy—a plane of ultimate reference, the ground-plane of the sidereal system.' We have now to consider what are the relations of the whole body of the stars to this Galactic Circle—this plane of ultimate reference for the whole stellar universe.

If we look at the heavens on a starry night, the whole vault appears to be thickly strewn with stars of various degrees of brightness, so that we could hardly say that any extensive region—the north, east, south, or west, or the portion vertically above us—is very conspicuously deficient or superior in numbers. In every part there are to be found a fair proportion of stars of the first two or three magnitudes, while where these may seem deficient a crowd of smaller stars takes their place.

But an accurate survey of the visible stars shows that there is a large amount of irregularity in their distribution, and that all magnitudes are really more numerous in or near the Milky Way, than at a distance from it, though not in so large a degree as to be very conspicuous to the naked eye. The area of the whole of the Milky Way cannot be estimated at more than one-seventh of the whole sphere, while some astronomers reckon it at only one-tenth. If stars of any particular size were uniformly distributed, at most one-seventh of the whole number should be found within its limits. But Mr. Gore finds that of 32 stars brighter than the second magnitude 12 lie upon the Milky Way, or considerably more than twice as many as there should be if they were uniformly distributed. And in the case of the 99 stars which are brighter than the third magnitude 33 lie upon the Milky Way, or one-third instead of one-seventh. Mr. Gore also counted all the stars in Heis's Atlas which lie upon the Milky Way, and finds there are 1186 out of a total of 5356, a proportion of between a fourth and a fifth instead of a seventh.

The late Mr. Proctor in 1871 laid down on a chart two feet diameter all the stars down to magnitude $9^1/_2$ given in Agrelander's forty large charts of the stars visible in the northern hemisphere. They were 324,198 in number, and they distinctly showed by their greater density not only the whole course of the Milky Way but also its more luminous portions and many of the curious dark rifts and vacuities, which latter are almost wholly avoided by these stars.

Later on Professor Seeliger of Munich made an investigation of the relation of more than 135,000 stars down to the ninth magnitude to the Milky Way, by dividing the whole of the heavens into nine regions, one and nine being circles of 20° wide (equal to 40° diameter) at the two poles of the Galaxy; the middle region, five, is a zone 20° wide including the Milky Way itself, and the other six intermediate zones are each 20° wide. The following table shows the results as given by Professor Newcomb, who has made some alterations in the last column of 'Density of Stars' in order to correct differences in the estimate of magnitudes by the different authorities.

Regions.	Area in Degree.	Number of Stars.	Density.
I.	1,398.7	4,277	2.78
II.	3,146.9	10,185	3.03
III.	5,126.6	19,488	3.54
IV.	4,589.8	24,492	5.32
V.	4,519.5	33,267	8.17
VI.	3,971.5	23,580	6.07
VII.	2,954.4	11,790	3.71
VIII.	1,796.6	6,375	3.21
IX.	468.2	1,644	3.14

N.B.—The inequality of the N. and S. areas is because the enumeration of the stars only went as far as 24° S. Decl., and therefore included only a part of Regions VII., VIII., and IX.

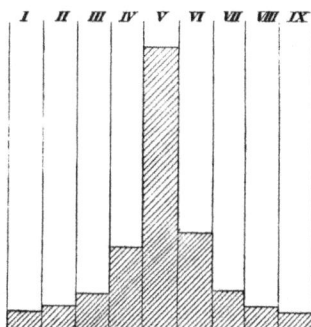

DIAGRAM OF STAR-DENSITY
From Herschel's Gauges
(as given by Professor Newcomb, p. 251).

Upon this table of densities Professor Newcomb remarks as follows:—'The star-density in the several regions increases continuously from each pole (regions I. and IX.) to the Galaxy itself (region V.). If the latter were a simple ring of stars surrounding a spherical system of stars, the star-density would be about the same in regions I., II., and III., and also in VII., VIII., and IX., but would suddenly increase in IV. and VI. as the boundary of the ring was approached. Instead of such being the case, the numbers 2.78, 3.03, and 3.54 in the north, and 3.14, 3.21, and 3.71 in the south, show a progressive increase from the galactic pole to the Galaxy itself. The conclusion to be drawn is a fundamental one. The universe, or at least the denser portion of it, is really flattened between the galactic poles, as supposed by

Herschel and Struve.'

But looking at the series of figures in the table, and again as quoted by Professor Newcomb, they seem to me to show in some measure what he says they do not show. I therefore drew out the above diagram from the figures in the table, and it certainly shows that the density in regions I., II., and III., and in regions VII., VIII., and IX., may be said to be 'about the same,' that is, they increase very slowly, and that they *do* 'suddenly increase' in IV. and VI. as the boundary of the Galaxy is approached. This may be explained either by a flattening towards the poles of the Galaxy, or by the thinning out of stars in that direction.

In order to show the enormous difference of star-density in the Galaxy and at the galactic poles, Professor Newcomb gives the following table of the Herschelian gauges, on which he only remarks that they show an enormously increased density in the galactic region due to the Herschels having counted so many more stars there than any other observers.

Region, .	I.	II.	III.	IV.	V.	VI.	VII.	VIII.	IX.
Density, .	107	154	281	560	2,019	672	261	154	111

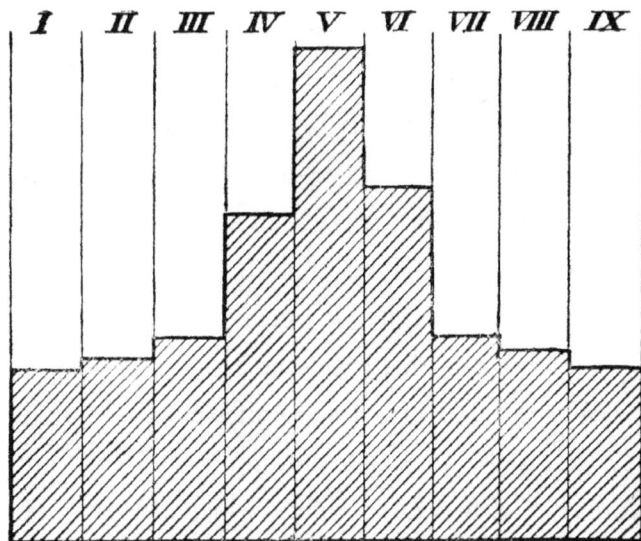

DIAGRAM OF STAR-DENSITY From a table in The Stars (p. 249).

But an important characteristic of these figures is, that the Herschels alone surveyed the whole of the heavens from the north to the south pole, that they did this with instruments of the same size and quality, and that from almost life-long experience in this particular work they were unrivalled in their power of counting rapidly and accurately the stars that passed over each field of view of their telescopes. Their results, therefore, must be held to have a comparative value far above those of any other observer or combination of observers. I have therefore thought it advisable to draw a diagram from their figures, and it will be seen how strikingly it agrees with the former diagram in the very slow increase of star-richness in the first three regions north and south, the sudden increase in regions IV. and VI. as we approach the Galaxy, while the only marked difference is in the enormously greater richness of the Galaxy itself, which is an undoubtedly real phenomenon, and is brought out here by the unrivalled observing power of the two greatest astronomers in this special department that have ever lived.

We shall find later on that Professor Newcomb himself, as the result of a quite different inquiry arrives at a result in accordance with these diagrams which will then be again referred to. As this is a very interesting subject, it will be well to give another diagram from two tables of star-density in Sir John Herschel's volume already quoted. The tables are as follows:—

Zones of Galactic South Polar Distance.	Average number of Stars per Field of 15'.
0° to 15°	6.05
15° to 30°	6.62
30° to 45°	9.08
45° to 60°	13.49
60° to 75°	26.29
75° to 90°	59.06

In these tables the Milky Way itself is taken as occupying two zones of 15° each, instead of one of 20° as in Professor Newcomb's tables, so that the excess in the number of stars over the other zones is not so large. They show also a slight preponderance in all the zones of the southern hemisphere, but this is not great, and may probably be due to the clearer atmosphere of the Cape of Good Hope as compared with that of England.

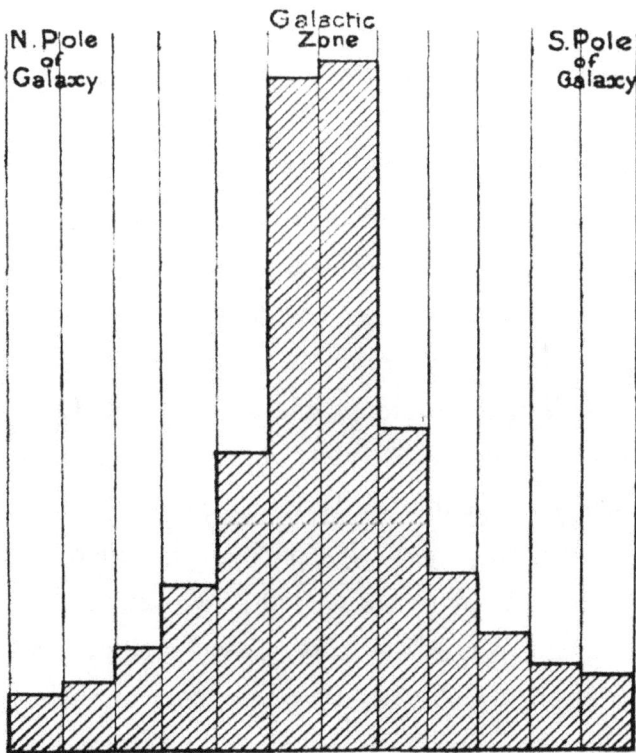

DIAGRAM OF STAR-DENSITY. From Table in Sir J. Herschel's Outlines of Astronomy (10th ed., pp. 577-578).

It need only be noted here that this diagram shows the same general features as those already given, of a continuous increase of star-density from the poles of the Galaxy, but more rapidly as the Galaxy itself is more nearly approached. This fact must, therefore, be accepted as indisputable.

Clusters and Nebulæ in Relation to the Galaxy

An important factor in the structure of the heavens is afforded by the distribution of the two classes of objects known as clusters and nebulæ. Although we can form an almost continuous series from double stars which revolve round their common centre of gravity, through triple and quadruple stars, to groups and aggregations of indefinite extent—of which the Pleiades form a good example, since the six stars visible to the naked eye are increased to hundreds by high telescopic powers, while photographs with three hours' exposure show more than 2000 stars—yet none of these correspond to the large class known as clusters, whether globular or irregular, which are very numerous, about 600 having been recorded by Sir John Herschel more than fifty years ago. Many of these are among the most beautiful and striking objects in the heavens even with a very small telescope or good opera-glass. Such is the luminous spot called Praesepe, or the Beehive in the constellation Cancer, and another in the sword handle of Perseus.

In the southern hemisphere there is a hazy star of about the fourth magnitude, Omega Centauri, which with a good telescope is seen to be really a magnificent cluster nearly two-thirds the diameter of the moon, and described by Sir John Herschel as very gradually increasing in brightness to the centre, and composed of innumerable stars of the thirteenth and fifteenth magnitudes, forming the richest and largest object of the kind in the heavens. He describes it as having rings like lace-work formed of the larger stars. By actual count, on a good photograph, there are more than 6000 stars, while other observers consider that there are at least 10,000. In the northern hemisphere one of the finest is that in the constellation Hercules, known as 13 Messier. It is just visible to the naked eye or with an opera glass as a hazy star of the sixth magnitude, but a good telescope shows it to be a globular cluster, and the great Lick telescope resolves even the densest central portion into distinct stars, of which Sir John Herschel considered there were many thousands. These two fine clusters are figured in many of the modern popular works on astronomy, and they afford an excellent idea of these beautiful and remarkable objects, which, when more thoroughly studied, will probably aid in elucidating some of the obscure problems connected with the constitution and development of the stellar universe.

But for the purpose of the present work the most

interesting fact connected with star-clusters is their remarkable distribution in the heavens. Their special abundance in and near the Milky Way had often been noted, but the full importance of the fact could not be appreciated till Mr. Proctor and, later, Mr. Sidney Waters marked down, on maps of the two hemispheres, all the star-clusters and nebulæ in the best catalogues. The result is most interesting. The clusters are seen to be thickly strewn over the entire course of the Milky Way, and along its margins, while in every other part of the heavens they are thinly scattered at very distant intervals, with the one exception of the Magellanic clouds of the southern hemisphere where they are again densely grouped; and if anything were needed to prove the physical connection of these clusters with the Galaxy it would be their occurrence in these extensive nebulous patches which seem like outlying portions of the Milky Way itself. With these two exceptions probably not one-twentieth part of the whole number of star-clusters are found in any part of the heavens remote from the Milky Way.

Nebulæ were for a long time confounded with star-clusters, because it was thought that with sufficient telescopic power they could all be resolvable into stars as in the case of the Milky Way itself. But when the spectroscope showed that many of the nebulæ consisted wholly or mainly of glowing gases, while neither the highest powers of the best telescopes nor the still greater powers of the photographic plate gave any indications of resolvability, although a few stars were often found to be, as it were, entangled in them, and evidently forming part of them, it was seen that they constituted a distinct stellar phenomenon, a view which was enforced and rendered certain by their quite unique mode of distribution. A few of the larger and irregular type, as in the case of the grand Orion nebula visible to the naked eye, the great spiral nebula in Andromeda, and the wonderful Keyhole nebula round Eta Argûs, are situated in or near the Milky Way; but with these and a few other exceptions the overwhelming majority of the smaller irresolvable nebulæ appear to avoid it, there being a space almost wholly free from nebulæ along its borders, both in the northern and southern hemispheres; while the great majority are spread over the sky, far away from it in the southern hemisphere, and in the north clustering in a very marked degree around the galactic pole. The distribution of nebulæ is thus seen to be the exact opposite to that of the star-clusters, while both are so distinctly related to the position of the Milky Way—the ground-plane of the sidereal system, as Sir John Herschel termed it—that we are compelled to include them all as connected portions of one grand and, to some extent, symmetrical universe, whose remarkable and opposite mode of distribution over the heavens may probably afford a clue to the mode of development of that universe and to the changes that are even now taking place within it. The maps referred to above are of such great importance, and are so essential to a clear comprehension of the nature and constitution of the vast sidereal system which surrounds us, that I have, with the permission of the Royal Astronomical Society, reproduced them here. (See end of volume.)

A careful examination of them will give a clearer idea of the very remarkable facts of distribution of star-clusters and nebulæ than can be afforded by any amount of description or of numerical statements.

The forms of many of the nebulæ are very curious. Some are quite irregular, as the Orion nebula, the Keyhole nebula in the southern hemisphere, and many others. Some show a decidedly spiral form, as those in Andromeda and Canes Venatici; others again are annular or ring-shaped, as those in Lyra and Cygnus, while a considerable number are termed planetary nebulæ, from their exhibiting a faint circular disc like that of a planet. Many have stars or groups of stars evidently forming parts of them, and this is especially the case with those of the largest size. But all these are comparatively few in number and more or less exceptional in type, the great majority being minute cloudy specks only visible with good telescopes, and so faint as to leave much doubt as to their exact shape and nature. Sir John Herschel catalogued 5000 in 1864, and more than 8000 were discovered up to 1890; while the application of the camera has so increased the numbers that it is thought there may really be many hundreds of thousands of them.

The spectroscope shows the larger irregular nebulæ to be gaseous, as are the annular and planetary nebulæ as well as many very brilliant white stars; and all these objects are most frequent in or near the Milky Way. Their spectra show a green line not produced by any terrestrial element. With the great Lick telescope several of the planetary nebulæ have been found to be irregular and sometimes to be formed of compressed or looped rings and other curious forms.

Many of the smaller nebulæ are double or triple, but whether they really form revolving systems is not yet known. The great mass of the small nebulæ that occupy large tracts of the heavens remote from the Galaxy are often termed irresolvable nebulæ, because the highest powers of the largest telescopes show no indication of their being star-clusters, while they are too faint to give any definite indications of structure in the spectroscope. But many of them resemble comets in their forms, and it is thought not impossible that they may be not very dissimilar in constitution.

We have now passed in review the main features presented to us in the heavens outside the solar system, so far as regards the numbers and distribution of the lucid stars (those visible to the naked eye) as well as those brought to view by the telescope; the form and chief characteristics of the Milky Way or Galaxy; and lastly, the numbers and distribution of those interesting objects—star-clusters and nebulæ in their special relations to the Milky Way. This examination has brought clearly before us the unity of the whole visible universe; that everything we can see, or obtain any knowledge of, with all the resources of modern gigantic telescopes, of the photographic plate, and of the even more marvellous spectroscope, forms parts of one vast system which may be shortly and appropriately termed the Stellar universe.

In our next chapter we shall carry the investigation a step further, by sketching in outline what is known of the motions and distances of the stars, and thus obtain some important information bearing upon our special subject of inquiry.

CHAPTER V

DISTANCE OF THE STARS—THE SUN'S MOTION

THROUGH SPACE

In early ages, before any approximate idea was reached of the great distances of the stars from us, the simple conception of a crystal sphere to which these luminous points were attached and carried round every day on an axis near which our pole-star is situated, satisfied the demands for an explanation of the phenomena. But when Copernicus set forth the true arrangement of the heavenly bodies, earth and planets alike revolving round the sun at distances of many millions of miles, and when this scheme was enforced by the laws of Kepler and the telescopic discoveries of Galileo, a difficulty arose which astronomers were unable satisfactorily to overcome. If, said they, the earth revolves round the sun at a distance which cannot be less (according to Kepler's measurement of the distance of Mars at opposition) than $13^1/_2$ millions of miles, then how is it that the nearer stars are not seen to shift their apparent places when viewed from opposite sides of this enormous orbit? Copernicus, and after him Kepler and Galileo, stoutly maintained that it was because the stars were at such an enormous distance from us that the earth's orbit was a mere point in comparison. But this seemed wholly incredible, even to the great observer Tycho Brahé, and hence the Copernican theory was not so generally accepted as it otherwise would have been.

Galileo always declared that the measurement would some day be made, and he even suggested the method of effecting it which is now found to be the most trustworthy. But the sun's distance had to be first measured with greater accuracy, and that was only done in the latter part of the eighteenth century by means of transits of Venus; and by later observations with more perfect instruments it is now pretty well fixed at about 92,780,000 miles, the limits of error being such that $92^3/_4$ millions may perhaps be quite as accurate.

With such an enormous base-line as twice this distance, which is available by making observations at intervals of about six months when the earth is at opposite points in its orbit, it seemed certain that some parallax or displacement of the nearer stars could be found, and many astronomers with the best instruments devoted themselves to the work. But the difficulties were enormous, and very few really satisfactory results were obtained till the latter half of the nineteenth century. About forty stars have now been measured with tolerable certainty, though of course with a considerable margin of possible or probable error; and about thirty more, which are found to have a parallax of one-tenth of a second or less, must be considered to leave a very large margin of uncertainty.

The two nearest fixed stars are Alpha Centauri and 61 Cygni. The former is one of the brightest stars in the southern hemisphere, and is about 275,000 times as far from us as the sun. The light from this star will take $4^1/_4$ years to reach us, and this 'light-journey,' as it is termed, is generally used by astronomers as an easily remembered mode of recording the distances of the fixed stars, the distance in miles—in this case about 25 millions of millions—being very cumbrous. The other star, 61 Cygni, is only of about the fifth magnitude, yet it is the second nearest to us, with a light-journey of about $7^1/_4$ years. If we had no other determinations of distance than these two, the facts would be of the highest importance. They teach us, first, that magnitude or brightness of a star is no proof of nearness to us, a fact of which there is much other evidence; and in the second place, they furnish us with a probable minimum distance of independent suns from one another, which, in proportion to their sizes, some being known to be many times larger than our sun, is not more than we might expect. This remoteness may be partly due to those which were once nearer together having coalesced under the influence of gravitation.

As this measurement of the distance of the nearer stars should be clearly understood by every one who wishes to obtain some real comprehension of the scale of this vast universe of which we form a part, the method now adopted and found to be most effectual will be briefly explained.

Everyone who is acquainted with the rudiments of trigonometry or mensuration, knows that an inaccessible distance can be accurately determined if we can measure a base-line from both ends of which the inaccessible object can be seen, and if we have a good instrument with which to measure angles. The accuracy will mainly depend upon our base-line being not excessively short in comparison with the distance to be measured. If it is as much as half or even a quarter as long the measurement may be as accurate as if directly performed over the ground, but if it is only one-hundredth or one-thousandth part as long, a very small error either in the length of the base or in the amount of the angles will produce a large error in the result.

In measuring the distance of the moon, the earth's diameter, or a considerable portion of it, has served as a base-line. Either two observers at great distances from each other, or the same observer after an interval of nine or ten hours, may examine the moon from positions six or seven thousand miles apart, and by accurate measurements of its angular distance from a star, or by the time of its passage over the meridian of the place as observed with a transit instrument, the angular displacement can be found and the distance determined with very great accuracy, although that distance is more than thirty times the length of the base. The distance of the planet Mars when nearest to us has been found in the same way. His distance from us even when at his nearest point during the most favourable oppositions is about 36 million miles, or more than four thousand times the earth's diameter, so that it requires the most delicate observations many times repeated and with the finest instruments to obtain a tolerably approximate result. When this is done, by Kepler's law of the fixed proportion between the distances of planets from the sun and their times of revolution, the proportionate distance of all the other planets and that of the sun can be ascertained. This method, however, is not sufficiently accurate to satisfy astronomers, because upon the sun's distance that of every other member of the solar system depends. Fortunately there are two other methods by which this important measurement has been made with much greater approach to certainty and precision.

DIAGRAM ILLUSTRATING THE TRANSIT OF VENUS.

The first of these methods is by means of the rare occasions when the planet Venus passes across the sun's disc as seen from the earth. When this takes place, observations of the transit, as it is termed, are made at remote parts of the earth, the distance between which places can of course easily be calculated from their latitudes and longitudes. The diagram here given illustrates the simplest mode of determining the sun's distance by this observation, and the following description from Proctor's *Old and New Astronomy* is so clear that I copy it verbally:—'V represents Venus passing between the Earth E and the Sun S; and we see how an observer at E will see Venus as at v', while an observer at E' will see her as at v. The measurement of the distance v v', as compared with the diameter of the sun's disc, determines the angle v V v' or E V E'; whence the distance E V can be calculated from the known length of the base-line E E'. For instance, it is known (from the known proportions of the Solar System as determined from the times of revolution by Kepler's third law) that E V bears to V v the proportion 28 to 72, or 7 to 18; whence E E' bears to v v' the same proportion. Suppose, now, that the distance between the two stations is known to be 7000 miles, so that v v' is 18,000 miles; and that v v' is found by accurate measurement to be $\frac{1}{48}$ part of the sun's diameter. Then the sun's diameter, as determined by this observation, is 48 times 18,000 miles, or 864,000 miles; whence from his known apparent size, which is that of a globe $107^{1}/_{3}$ times farther away from us than its own diameter, his distance is found to be 92,736,000 miles.'

Of course, there being two observers, the proportion of the distance v v' to the diameter of the sun's disc cannot be measured directly, but each of them can measure the apparent angular distance of the planet from the sun's upper and lower margins as it passes across the disc, and thus the angular distance between the two lines of transit can be obtained. The distance v v' can also be found by accurately noting the times of the upper and lower passage of Venus, which, as the line of transit is considerably shorter in one than the other, gives by the known properties of the circle the exact proportion of the distance between them to the sun's diameter; and as this is found to be the most accurate method, it is the one generally adopted. For this purpose the stations of the observers are so chosen that the length of the two chords, v and v', may have a considerable difference, thus rendering the measurement more easy.

The other method of determining the sun's distance is by the direct measurement of the velocity of light. This was first done by the French physicist, Fizeau, in 1849, by the use of rapidly revolving mirrors, as described in most works on physics. This method has now been brought to such a decree of perfection that the sun's distance so determined is considered to be equally trustworthy with that derived from the transits of Venus. The reason that the determination of the velocity of light leads to a determination of the sun's distance is, because the time taken by light to pass from the sun to the earth is independently known to be 8 min. $13^{1}/_{3}$ sec. This was discovered so long ago as 1675 by means of the eclipses of Jupiter's satellites. These satellites revolve round the planet in from $1^{3}/_{4}$ to 16 days, and, owing to their moving very nearly in the plane of the ecliptic and the shadow of Jupiter being so large, the three which are nearest to the planet are eclipsed at every revolution. This rapid revolution of the satellites and frequency of the eclipses enabled their periods of recurrence to be determined with extreme accuracy, especially after many years of careful observation. It was then found that when Jupiter was at its farthest distance from the earth the eclipses of the satellites took place a little more than eight minutes later than the time calculated from the mean period of revolution, and when the planet was nearest to us the eclipses occurred the same amount earlier. And when further observation showed that there was no difference between calculation and observation when the planet was at its mean distance from us, and that the error arose and increased exactly in proportion to our varying distance from it, then it became clear that the only cause adequate to produce such an effect was, that light had not an infinite velocity but travelled at a certain fixed rate. This however, though a highly probable explanation, was not absolutely proved till nearly two centuries later, by means of two very difficult measurements—that of the actual distance of the sun from the earth, and that of the actual speed of light in miles per second; the latter corresponding almost exactly with the speed deduced from the eclipses of Jupiter's satellites and the sun's distance as measured by the transits of Venus.

(A) 5³⁄₄ INCHES FROM IT (ACCURATELY 5.72957795 INCHES)

But this problem of measuring the sun's distance, and through it the dimensions of the orbits of all the planets of our system, sinks into insignificance when compared with the enormous difficulties in the way of the determination of the distance of the stars. As a great many people, perhaps the majority of the readers of any popular scientific book, have little knowledge of mathematics and cannot realise what an angle of a minute or a second really means, a little explanation and illustration of these terms will not be out of place. An angle of one degree (1°) is the 360th part of a circle (viewed from its centre), the 90th part of a right angle, the 60th part of either of the angles of an equilateral triangle. To see exactly how much is an angle of one degree we draw a short line (B C) one-tenth of an inch long, and from a point we draw straight lines to B and C. Then the angle at A is one degree.

Now, in all astronomical work, one degree is considered to be quite a large angle. Even before the invention of the telescope the old observers fixed the position of the stars and planets to half or a quarter of a degree, while Mr. Proctor thinks that Tycho Brahé's positions of the stars and planets were correct to about one or two minutes of arc. But a minute of arc is obtained by dividing the line B C into sixty equal parts and seeing the distance between two of these with the naked eye from the point A. But as very long-sighted people can see very minute objects at 10 or 12 inches distance, we may double the distance A B, and then making the line B C one three-hundredth part of an inch long, we shall have the angle of one minute which Tycho Brahé was perhaps able to measure. How very large an amount a minute is to the modern astronomer is, however, well shown by the fact that the maximum difference between the calculated and observed positions of Uranus, which led Adams and Leverrier to search for and discover Neptune, was only $1^1/_2$ minutes, a space so small as to be almost invisible to the average eye, so that if there had been two planets, one in the calculated, the other in the observed place, they would have appeared as one to unassisted vision.

In order now to realise what one second of arc really means, let us look at the circle here shown, which is as nearly as possible one-tenth of an inch in diameter—(one-O-tenth of an inch). If we remove this circle to a distance of 28 feet 8 inches it will subtend an angle of one minute, and we shall have to place it at a distance of nearly 1730 feet—almost one-third of a mile—to reduce the angle to one second. But the very nearest to us of the fixed stars, Alpha Centauri, has a parallax of only three-fourths of a second; that is, the distance of the earth from the sun—about $92^3/_4$ millions of miles—would appear no wider, seen from the nearest star, than does three-fourths of the above small circle at one-third of a mile distance. To see this circle at all at that distance would require a very good telescope with a power of at least 100, while to see any small part of it and to measure the proportion of that part to the whole would need very brilliant illumination and a large and powerful astronomical telescope.

What is a Million?

But when we have to deal with millions, and even with hundreds and thousands of millions, there is another difficulty—that few people can form any clear conception of what a million is. It has been suggested that in every large school the walls of one room or hall should be devoted to showing a million at one view. For this purpose it would be necessary to have a hundred large sheets of paper each about 4 feet 6 inches square, ruled in quarter inch squares. In each alternate square a round black wafer or circle should be placed a little over-lapping the square, thus leaving an equal amount of white space between the black spots. At each tenth spot a double width should be left so as to separate each hundred spots (10 × 10). Each sheet would then hold ten thousand spots, which would all be distinctly visible from the middle of a room 20 feet wide, each horizontal or vertical row containing a thousand. One hundred such sheets would contain a million spots, and they would occupy a space 450 feet long in one row, or 90 feet long in five rows, so that they would entirely cover the walls of a room, about 30 feet square and 25 feet high, from floor to ceiling, allowing space for doors but not for windows, the hall or gallery being lighted from above. Such a hall would be in the highest degree educational in a country where millions are spoken of so glibly and wasted so recklessly; while no one can really appreciate modern science, dealing as it does with the unimaginably great and little, unless he is enabled to realise by actual vision, and summing up, what a vast number is comprised in *one* of those millions, which, in modern astronomy and physics, he has to deal with not singly only, but by hundreds and thousands or even by millions. In every considerable town, at all events, a hall or gallery should have a *million* thus shown upon its walls. It would in no way interfere with the walls being covered when required with maps, or ornamental hangings, or pictures; but when these were removed, the visible and countable million would remain as a permanent lesson to all visitors; and I believe that it would have widespread beneficial effects in almost every department of human thought and action. On a small scale any one can do this for himself by getting a hundred sheets of engineer's paper ruled in small squares, and making the spots very small; and even this would be impressive, but not so much so as on the larger scale.

In order to enable every reader of this volume at once to form some conception of the number of units in a million, I have made an estimate of the number of *letters* contained in it, and I find them to amount to about 420,000—considerably less than half a million. Try and realise, when reading it, that if every letter were a pound sterling, we waste as many pounds as there are letters in *two* such volumes whenever we build a battleship.

Having thus obtained some real conception of the immensity of a million, we can better realise what it must be to have every one of the dots above described, or every one of the letters in two such volumes as this lengthened out so as to be each a mile long, and even then we should have reached little more than a hundredth part of the distance from our earth to the sun. When, by careful consideration of these figures, we have even partially realised this enormous distance, we may take the next step, which is, to compare this distance with that of the nearest fixed star. We have seen that the parallax of that star is three-fourths of a second, an amount which implies that the star is 271,400 times as far from us as our sun is. If after *seeing* what a million is, and knowing that the sun is $92^3/_4$ times this distance from us in miles—a distance which itself is almost inconceivable to us— we find that we have to multiply this almost inconceivable distance 271,400 times—more than a quarter of a million times—to reach the *nearest* of the fixed stars, we shall begin to realise, however imperfectly, how vast is the system of suns around us, and on what a scale of immensity the material universe, which we see so gloriously displayed in the starry heavens and the mysterious galaxy, is constructed.

This somewhat lengthy preliminary discussion is thought necessary in order that my readers may form some idea of the enormous difficulty of obtaining any measurement whatever of such distances. I now propose to point out what the special difficulties are, and how they have been overcome; and thus I hope to be able to satisfy them that the figures astronomers give us of the distances of the stars are in no way mere guesses or probabilities, but are real measurements which, within certain not very wide limits of error, may be trusted as giving us correct ideas of the magnitude of the visible universe.

Measurement of Stellar Distances

The fundamental difficulty of this measurement is, of course, that the distances are so vast that the longest available base-line, the diameter of the earth's orbit, only subtends an angle of little more than a second from the nearest star, while for all the rest it is less than one second and often only a small fraction of it. But this difficulty, great as it is, is rendered far greater by the fact that there is no fixed point in the heavens from which to measure, since many of the stars are known to be in motion, and all are believed to be so in varying degrees, while the sun itself is now known to be moving among the stars at a rate which is not yet accurately determined, but in a direction which is fairly well known. As the various motions of the earth while passing round the sun, though extremely complex, are very accurately known, it was first attempted to determine the changed position of stars by observations, many times repeated at six months' intervals, of the moment of their passage over the meridian and their distance from the zenith; and then by allowing for all the known motions of the earth, such as precession of the equinoxes and nutation of the earth's axis, as well as for refraction and for the aberration of light, to determine what residual effect was due to the difference of position from which the star was viewed; and a result was thus obtained in several cases, though almost always a larger one than has been found by later observations and by better methods. These earlier observations, however perfect the instruments and however skilful the observer, are liable to errors which it seems impossible to avoid. The instruments themselves are subject in all their parts to expansion and contraction by changes of temperature; and when these changes are sudden, one part of the instrument may be affected more than another, and this will often lead to minute errors which may seriously affect the amount to be measured when that is so small. Another source of error is due to atmospheric refraction, which is subject to changes both from hour to hour and at different seasons. But perhaps most important of all are minute changes in level of the foundations of the instruments even when they are carried down to solid rock. Both changes of temperature and changes of moisture of the soil produce minute alterations of level; while earth-tremors and slow movements of elevation or depression are now known to be very frequent. Owing to all these causes, actual measurements of differences of position at different times of the year, amounting to small fractions of a second, are found to be too uncertain for the determination of such minute angles with the required accuracy.

But there is another method which avoids almost all these sources of error, and this is now generally preferred and adopted for these measurements. It is, that of measuring the distance between two stars situated apparently very near each other, one of which has large proper motion, while the other has none which is measurable. The proper motions of the stars was first suspected by Halley in 1717, from finding that several stars, whose places had been given by Hipparchus, 130 B.C., were not in the positions where they now ought to be; and other observations by the old astronomers, especially those of occultations of stars by the moon, led to the same result. Since the time of Halley very accurate observations of the stars have been made, and in many cases it is found that they move perceptibly from year to year, while others move so slowly that it is only after forty or fifty years that the motion can be detected. The greatest proper motions yet determined amount to between 7" and 8" in a year, while other stars require twenty, or even fifty or a hundred years to show an equal amount of displacement. At first it was thought that the brightest stars would have the largest proper motion, because it was supposed they were nearest to us, but it was soon found that many small and quite inconspicuous stars moved as rapidly as the most brilliant, while in many very bright stars no proper motion at all can be detected. That which moves most rapidly is a small star of less than the sixth magnitude.

It is a matter of common observation that the motion of things at a distance cannot be perceived so well as when near, even though the speed may be the same. If a man is seen on the top of a hill several miles off, we have to observe him closely for some time before we can be sure whether he is walking or standing still. But objects so enormously distant as we now know that the stars are, may be moving at the rate of many miles in a second and yet require years of observation to detect any movement at all.

The proper motions of nearly a hundred stars have now been ascertained to be more than one second of arc annually, while a large number have less than this, and the majority have no perceptible motion, presumably due to their enormous distance from us. It is therefore not difficult in most cases to find one or two motionless stars sufficiently close to a star having a large proper motion (anything more than one-tenth of a second is so called) to serve as fixed points of measurement. All that is then required is, to measure with extreme accuracy the angular distance of the moving from the fixed stars at intervals of six months. The measurements can be made, however, on every fine night, each one being compared with one at nearly an interval of six months from it. In this way a hundred or more measurements of the same star may be made in a year, and the mean of the whole, allowance being made for proper motion in the interval, will give a much more accurate result than any single measurement. This kind of measurement can be made with extreme accuracy when the two stars can be seen together in the field of the telescope; either by the use of a micrometer, or by means of an instrument called a heliometer, now often constructed for the purpose. This is an astronomical telescope of rather large size, the object glass of which is cut in two straight across the centre, and the two halves made to slide upon each other by means of an exceedingly fine and accurate screw-motion, so adjusted and tested as to measure the angular distance of two objects with extreme accuracy. This is done by the number of turns of the screw required to bring the two stars into contact with each other, the image of each one being formed by one of the halves of the object glass.

But the greatest advantage of this method of determining parallax is, as Sir John Herschell points out, that it gets rid of all the sources of error which render the older methods so uncertain and inaccurate. No corrections are required for precession, nutation, or aberration, since these affect both stars alike, as is the case also with refraction; while alterations of level of the instrument have no prejudicial effect, since the measures of angular distance taken by this method are quite independent of such movements. A test of the accuracy of the determination of parallax by this instrument is the very close agreement of different observers, and also their agreement with the new and perhaps even superior method by photography. This method was first adopted by Professor Pritchard of the Oxford Observatory, with a fine reflector of thirteen inches aperture. Its great advantage is, that all the small stars in the vicinity of the star whose parallax is sought are shown in their exact positions upon the plate, and the distances of all of them from it can be very accurately measured, and by comparing plates taken at six months' intervals, each of these stars gives a determination of parallax, so that the mean of the whole will lead to a very accurate result. Should, however, the result from any one of these stars differ considerably from that derived from the rest, it will be due in all probability to that star having a proper motion of its own, and it may therefore be rejected. To illustrate the amount of labour bestowed by astronomers on this difficult problem, it may be mentioned that for the photographic measurement of the star 61 Cygni, 330 separate plates were taken in 1886-7, and on these 30,000 measurements of distances of the pairs of star-images were made. The result agreed closely with the best previous determination by Sir Robert Ball, using the micrometer, and the method was at once admitted by astronomers as being of the greatest value.

Although, as a rule, stars having large proper motions are found to be comparatively near us, there is no regular proportion between these quantities, indicating that the rapidity of the motion of the stars varies greatly. Among fifty stars whose distances have been fairly well determined, the rate of actual motion varies from one or two up to more than a hundred miles per second. Among six stars with less than a tenth of a second of annual proper motion there is one with a parallax of nearly half a second, and another of one-ninth of a second, so that they are nearer to us than many stars which move several seconds a year. This may be due to actual slowness of motion, but is almost certainly caused in part by their motion being either towards us or away from us, and therefore only measurable by the spectroscope; and this had not been done when the lists of parallaxes and proper motions from which these facts are taken were published. It is evident that the actual direction and rate of motion of a star cannot be known till this radial movement, as it is termed—that is, towards or away from us—has been measured; but as this element always tends to increase the visually observed rate of motion, we cannot, through its absence, exaggerate the actual motions of the stars.

The Sun's Movement Through Space

But there is yet another important factor which affects the apparent motions of all the stars—the movement of our sun, which, being a star itself, has a proper motion of its own. This motion was suspected and sought for by Sir William Herschel a century ago, and he actually determined the direction of its motion towards a point in the constellation Hercules, not very far removed from that fixed upon as the average of the best observations since made. The method of determining this motion is very simple, but at the same time very difficult. When we are travelling in a railway carriage near objects pass rapidly out of sight behind us, while those farther from us remain longer in view, and very distant objects appear almost stationary for a considerable time. For the same reason, if our sun is moving in any direction through space, the nearer stars will appear to travel in an opposite direction to our movement, while the more distant will remain quite stationary. This movement of the nearest stars is detected by an examination and comparison of their proper motions, by which it is found that in one part of the heavens there is a preponderance of the proper motions in one direction and a deficiency in the opposite direction, while in the directions at right angles to these the proper motions are not on the average greater in one direction than in the opposite. But the proper motions of the stars being themselves so minute, and also so irregular, it is only by a most elaborate mathematical investigation of the motions of hundreds or even of thousands of stars, that the direction of the solar motion can be determined. Till quite recently astronomers were agreed that the motion was towards a point in Hercules near the outstretched arm in the figure of that constellation. But the latest inquiries into this problem, involving the comparison of the motions of several thousand stars in all parts of the heavens, have led to the conclusion that the most probable direction of the 'solar apex' (as the point towards which the sun is moving is termed), is in the adjacent constellation Lyra, and not far from the brilliant star Vega. This is the position which Professor Newcomb of Washington thinks most probable, though there is still room for further investigation. To determine the rate of the motion is very much more difficult than to fix its direction, because the distances of so few stars have been determined, and very few indeed of these lie in the directions best adapted to give accurate results. The best measurements down to 1890 led to a motion of about 15 miles a second. But more recently the American astronomer, Campbell, has determined by the spectroscope the motion in the line of sight of a considerable number of stars towards and away from the solar apex, and by comparing the average of these motions, he derives a motion for the sun of about $12^{1}/_{2}$ miles a second, and this is probably as near as we can yet reach towards the true amount.

Some Numerical Results of the Above Measurements

The measurements of distances and proper motions of a considerable number of the stars, of the motion of our sun in space (its proper motion), together with accurate determinations of the comparative brilliancy of the brightest stars as compared with our sun and with each other, have led to some very remarkable numerical results which serve as indications of the scale of magnitude of the stellar universe.

The parallaxes of about fifty stars have now been repeatedly measured with such consistent results that Professor Newcomb considers them to be fairly trustworthy, and these vary from one-hundredth to three-quarters of a second. Three more, all stars of the first magnitude—Rigel, Canopus, and Alpha Cygni—have no measurable parallax, notwithstanding the long-continued efforts of many astronomers, affording a striking example of the fact that brilliancy alone is no test of proximity. Six more stars have a parallax of only one-fiftieth of a second, and five of these are either of the first or second magnitudes. Of these nine stars having very small parallax or none, six are situated in or near to the Milky Way, another indication of exceeding remoteness, which is further shown by the fact that they all have a very small proper motion or none at all. These facts support the conclusion, which had been already reached by astronomers from a careful study of the distribution of the stars, that the larger portion of the stars of all magnitudes scattered throughout the Milky Way or along its borders really belong to the same great system, and may be said to form a part of it. This is a conclusion of extreme importance because it teaches us that the grandest of the suns, such as Rigel and Betelgeuse in the constellation Orion, Antares in the Scorpion, Deneb in the Swan (Alpha Cygni), and

Canopus (Alpha Argus), are in all probability as far removed from us as are the innumerable minute stars which give the nebulous or milky appearance to the Galaxy.

It is well to consider for a moment what these facts mean. Professor S. Newcomb, one of the highest authorities on these problems, tells us that the long series of measurements to discover the parallax of Canopus, the brightest star in the southern hemisphere, would have shown a parallax of one-hundredth of a second, had such existed. Yet the results always seemed to converge to a mean of 0".000! Suppose, then, we assume the parallax of this star to be somewhat less than the hundredth of a second—let us say $^1/_{125}$ of a second. At the distance this gives, light would take almost exactly 400 years to reach us, so that if we suppose this very brilliant star to be situated a little on this side of the Galaxy, we must give to that great luminous circle of stars a distance of about 500 light years. We shall now perceive the advantage of being able to realise what a million really is. A person who had once seen a wall-space more than 100 feet long and 20 feet high completely covered with quarter-inch spots a quarter of an inch apart; and then tried to imagine every spot to be a mile long and to be placed end to end in one row, would form a very different conception of a million miles than those who almost daily *read* of millions, but are quite unable to visualise even one of them. Having really seen one million, we can partially realise the velocity of light, which travels over this million miles in a little less than $5^1/_2$ seconds; and yet light takes more than $4^1/_3$ years at this inconceivable speed to come to us from the very *nearest* of the stars. To realise this still more impressively, let us take the *distance* of this nearest star, which is 26 *millions* of *millions* of miles. Let us look in imagination at this large and lofty hall covered from floor to ceiling with quarter-inch spots—only *one* million. Let all these be imagined as miles. Then repeat this number of miles in a straight line, one after the other, as many times as there are spots in this hall; and even then you have reached only one twenty-sixth part of the distance to the nearest fixed star! This *million* times a *million* miles has to be repeated twenty-six times to reach the *nearest* fixed star; and it seems probable that this gives us a good indication of the distance from each other of at least all the stars down to the sixth magnitude, perhaps even of a large number of the telescopic stars. But as we have found that the bright stars of the Milky Way must be at least one hundred times farther from us than these nearest stars, we have found what may be termed a minimum distance for that vast star-ring. It may be immensely farther, but it is hardly possible that it should be anything less.

The Probable Size of the Stars

Having thus obtained an inferior limit for the distance of several stars of the first magnitude, and their actual brilliancy or light-emission as compared with our sun having been carefully measured, we have afforded us some indication of size though perhaps an uncertain one. By these means it has been found that Rigel gives out about ten thousand times as much light as our sun, so that if its surface is of the same brightness, it must be a hundred times the diameter of the sun. But as it is one of the white or Sirian type of stars it is probably very much more luminous, but even if it were twenty times brighter it would still have to be twenty-two and a half times the diameter of the sun; and as the stars of this type are probably wholly gaseous and much less dense than our sun, this enormous size may not be far from the truth. It is believed that the Sirian stars generally have a greater surface brilliancy than our sun. Beta Aurigæ, a star of the second magnitude but of the Sirian type, is one of the double stars whose distance has been measured, and this has enabled Mr. Gore to find the mass of the binary system to be five times that of the sun, and their light one hundred and seventeen times greater. Even if the density is much less than the sun's, the intrinsic brilliancy of the surface will be considerably greater. Another double star, Gamma Leonis, has been found to be three hundred times more brilliant than the sun if of the same density, but it would require to be seven times rarer than air to have the extent of surface needed to give the same amount of light if its surface emitted no more light than our sun from equal areas.

It is clear, therefore, that many of the stars are much larger than our sun as well as more luminous; but there are also large numbers of small stars whose large proper motions, as well as the actual measurement of some, prove them to be comparatively near to us which yet are only about one-fiftieth part as bright as the sun. These must, therefore, be either comparatively small, or if large must be but slightly luminous. In the case of some double stars it has been proved that the latter is the case; but it seems probable that others are very much smaller than the average. Up to the present time no means of determining the size of a star by actual measurement has been discovered, since their distances are so enormous that the most powerful telescopes show only a point of light. But now that we have really measured the distance of a good many stars we are able to determine an upper limit for their actual dimensions. As the nearest fixed star, Alpha Centauri, has a parallax of 0".75, this means that if this star has a diameter as great as our distance from the sun (which is not much more than a hundred times the sun's diameter) it would be seen to have a distinct disc about as large as that of Jupiter's first satellite. If it were even one-tenth of the size supposed it would probably be seen as a disc in our best modern telescopes. The late Mr. Ranyard remarks that if the Nebular Hypothesis is true, and our sun once extended as far as the orbit of Neptune, then, among the millions of visible suns there ought to be some now to be found in every stage of development. But any sun having a diameter at all approaching this size, and situated as far off as a hundred times the distance of Alpha Centauri, would be seen by the Lick telescope to have a disc half a second in diameter. Hence the fact that there are no stars with visible discs proves that there are no suns of the required size, and adds another argument, though not perhaps a strong one, against the acceptance of the Nebular Hypothesis.

CHAPTER VI

THE UNITY AND EVOLUTION OF THE STAR SYSTEM

The very condensed sketch now given of such of the discoveries of recent Astronomy as relate to the subject we are discussing will, it is hoped, give some idea both of the work already done and of the number of interesting problems yet remaining to be solved. The most eminent astronomers in every part of the world look forward to the solution of these problems not, perhaps, as of any great value in themselves, but as steps towards a more complete knowledge of our universe as a whole. Their aim is to do for the star-system what Darwin did for the organic world, to discover the processes of change that are at work in the heavens, and to learn how the mysterious nebulæ, the various types of stars, and the clusters and systems of stars are related to each other. As Darwin solved the problem of the origin of organic species from other species, and thus enabled us to understand how the whole of the existing forms of life have been developed out of pre-existing forms,One so astronomers hope to be able to solve the problem of the evolution of suns from some earlier stellar types, so as to be able, ultimately, to form some

124

intelligible conception of how the whole stellar universe has come to be what it is. Volumes have already been written on this subject, and many ingenious suggestions and hypotheses have been advanced. But the difficulties are very great; the facts to be co-ordinated are excessively numerous, and they are necessarily only a fragment of an unknown whole. Yet certain definite conclusions have been reached; and the agreement of many independent observers and thinkers on the fundamental principles of stellar evolution seems to assure us that we are progressing, if slowly yet with some established basis of truth, towards the solution of this, the most stupendous scientific problem with which the human intellect has ever attempted to grapple.

The Unity of the Stellar Universe

During the latter half of the nineteenth century the opinion of astronomers has been tending more and more to the conception that the whole of the visible universe of stars and nebulæ constitutes one complete and closely-related system; and during the last thirty years especially the vast body of facts accumulated by stellar research has so firmly established this view that it is now hardly questioned by any competent authority.

The idea that the nebulæ were far more remote from us

than the stars long held sway, even after it had been given up by its chief supporter. When Sir William Herschel, by means of his then unapproached telescopic power, resolved the Milky Way more or less completely into stars, and showed that numerous objects which had been classed as nebulæ were really clusters of stars, it was natural to suppose that those which still retained their cloudy appearance under the highest telescopic powers were also clusters or systems of stars, which only needed still higher powers to show their true nature. This idea was supported by the fact that several nebulæ were found to be more or less ring-shaped, thus corresponding on a smaller scale to the form of the Milky Way; so that when Herschel discovered thousands of telescopic nebulæ, he was accustomed to speak of them as so many distinct universes scattered through the immeasurable depths of space.

Now, although any real conception of the immensity of the one stellar universe, of which the Milky Way with its associated stars is the fundamental feature, is, as I have shown, almost unattainable, the idea of an unlimited number of other universes, almost infinitely remote from our own and yet distinctly visible in the heavens, so seized upon the imagination that it became almost a commonplace of popular astronomy and was not easily given up even by astronomers themselves. And this was in a large part due to the fact that Sir William Herschel's voluminous writings, being almost all in the Philosophical Transactions of the Royal Society, were very little read, and that he only indicated his change of view by a few brief sentences which might easily be overlooked. The late Mr. Proctor appears to have been the first astronomer to make a thorough study of the whole of Herschel's papers, and he tells us that he read them all over five times before he was able thoroughly to grasp the writer's views at different periods.

But the first person to point out the real teaching of the facts as to the distribution of the nebulæ was not an astronomer, but our greatest philosophical student of science in general, Herbert Spencer. In a remarkable essay on 'The Nebular Hypothesis' in the *Westminster Review* of July, 1858, he maintained that the nebulæ really formed a part of our own Galaxy and of our own stellar universe. A single passage from his paper will indicate his line of argument, which, it may be added, had already been partially set forth by Sir John Herschel in his *Outlines of Astronomy*.

'If there were but one nebula, it would be a curious coincidence were this one nebula so placed in the distant regions of space as to agree in direction with a starless spot in our own sidereal system. If there were but two nebulæ, and both were so placed, the coincidence would be excessively strange. What, then, shall we say on finding that there are thousands of nebulæ so placed? Shall we believe that in thousands of cases these far-removed galaxies happen to agree in their visible positions with the thin places in our own galaxy? Such a belief is impossible.'

He then applies the same argument to the distribution of the nebulæ as a whole:—'In that zone of celestial space where stars are excessively abundant, nebulæ are rare, while in the two opposite celestial spaces that are farthest removed from this zone, nebulæ are abundant. Scarcely any nebulæ lie near the galactic circle (or plane of the Milky Way); and the great mass of them lie round the galactic poles. Can this also be mere coincidence?' And he concludes, from the whole mass of the evidence, that 'the proofs of a physical connection become overwhelming.'

Nothing could be more clear or more forcible; but Spencer not being an astronomer, and writing in a comparatively little read periodical, the astronomical world hardly noticed him; and it was from ten to fifteen years later, when Mr. R.A. Proctor, by his laborious charts and his various papers read before the Royal and Royal Astronomical Societies from 1869 to 1875, compelled the attention of the scientific world, and thus did more perhaps than any other man to establish firmly the grand and far-reaching principle of the essential unity of the stellar universe, which is now accepted by almost every astronomical writer of eminence in the civilised world.

The Evolution of the Stellar Universe

Amid the enormous mass of observations and of suggestive speculation upon this great and most interesting problem, it is difficult to select what is most important and most trustworthy. But the attempt must be made, because, unless my readers have some knowledge of the most important facts bearing upon it (besides those already set forth), and also learn something of the difficulties that meet the inquirer into causes at every step of his way, and of the various ideas and suggestions which have been put forth to account for the facts and to overcome the difficulties, they will not be in a position to estimate, however imperfectly, the grandeur, the marvel, and the mystery of the vast and highly complex universe in which we live and of which we are an important, perhaps the most important, if not the only permanent outcome.

The Sun a Typical Star

It being now a recognised fact that the stars are suns, some knowledge of our own sun is an essential preliminary to an inquiry into their nature, and into the probable changes they have undergone.

The fact that the sun's density is only one-fourth that

of the earth, or less than one and a half times that of water, demonstrates that it cannot be solid, since the force of gravity at its surface being twenty-six and a half times that at the earth's surface, the materials of a solid globe would be so compressed that the resulting density would be at least twenty times greater instead of four times less than that of the earth. All the evidence goes to show that the body of the sun is really gaseous, but so compressed by its gravitative force as to behave more like a liquid. A few figures as to the vast dimensions of the sun and the amount of light and heat emitted by it will enable us better to understand the phenomena it presents, and the interpretation of those phenomena.

Proctor estimated that each square inch of the sun's surface emitted as much light as twenty-five electric arcs; and Professor Langley has shown by experiment that the sun is 5300 times brighter, and eighty-seven times hotter than the white-hot metal in a Bessemer converter. The actual amount of solar heat received by the earth is sufficient, if wholly utilised, to keep a three-horse-power engine continually at work on every square yard of the surface of our globe. The size of the sun is such, that if the earth were at its centre, not only would there be ample space for the moon's orbit, but sufficient for another satellite 190,000 miles beyond the moon, all revolving inside the sun. The mass of matter in the sun is 745 times greater than that of all the planets combined; hence the powerful gravitative force by which they are retained in their distant orbits.

What we see as the sun's surface is the photosphere or outer layer of gaseous or partially liquid matter kept at a definite level by the power of gravitation. The photosphere has a granular texture implying some diversity of surface or of luminosity; although the even contour of the sun's margin shows that these irregularities are not on a very large scale. This surface is apparently rent asunder by what are termed sun-spots, which were long supposed to be cavities, showing a dark interior; but are now thought to be due to downpours of cooled materials driven out from the sun, and forming the prominences seen during solar eclipses. They appear to be black, but around their margin is a shaded border or penumbra formed of elongated shining patches crossing and over-lapping, something like heaps of straw. Sometimes brilliant portions overhang the dark spots, and often completely bridge them over; and similar patches, called faculæ, accompany spots, and in some cases almost surround them.

Sun-spots are sometimes numerous on the sun's disc, sometimes very few, and they are of such enormous size that when present they can easily be seen with the naked eye, protected by a piece of smoked glass; or, better still, with an ordinary opera-glass similarly protected. They are found to increase in number for several years, and then to decrease; the maxima recurring after an average period of eleven years, but with no exactness, since the interval between two maxima or minima is sometimes only nine and sometimes as much as thirteen years; while the minima do not occur midway between two maxima, but much nearer to the succeeding than to the preceding one. What is more interesting is, that variations in terrestrial magnetism follow them with great accuracy; while violent commotions in the sun, indicated by the sudden appearance of faculæ, sun-spots, or prominences on the sun's limb, are always accompanied by magnetic disturbances on the earth.

What Surrounds the Sun

It has been well said that what we commonly term the sun is really the bright spherical nucleus of a nebulous body. This nucleus consists of matter in the gaseous state, but so compressed as to resemble a liquid or even a viscous fluid. About forty of the elements have been detected in the sun by means of the dark lines in its spectrum, but it is almost certain that all the elements, in some form or other, exist there. This semi-liquid glowing surface is termed the photosphere, since from it are given out the light and heat which reach our earth.

Immediately above this luminous surface is what is

termed the 'reversing layer' or absorbing layer, consisting of dense metallic vapours only a few hundred miles thick, and, though glowing, somewhat cooler than the surface of the photosphere. Its spectrum, taken, at the moment when the sun is totally darkened, through a slit which is directed tangentially to the sun's limb, shows a mass of bright lines corresponding in a large degree to the dark lines in the ordinary solar spectrum. It is thus shown to be a vaporous stratum which absorbs the special rays emitted by each element and forming its characteristic coloured lines, changing them into black lines. But as coloured lines are not found in this layer corresponding to all the black lines in the solar spectrum, it is now held that special absorption must also occur in the chromosphere and perhaps in the corona itself. Sir Norman Lockyer, in his volume on *Inorganic Evolution*, even goes so far as to say, that the true 'reversing layer' of the sun—that which by its absorption produced the dark lines in the solar spectrum—is now shown to be *not* the chromosphere itself but a layer above it, of lower temperature.

Above the reversing layer comes the chromosphere, a vast mass of rosy or scarlet emanations surrounding the sun to a depth of about 4000 miles. When seen during eclipses it shows a serrated waving outline, but subject to great changes of form, producing the prominences already mentioned. These are of two kinds: the 'quiescent,' which are something like clouds of enormous extent, and which keep their forms for a considerable time; and the 'eruptive,' which shoot out in towering tree-like flames or geyser-like eruptions, and while doing so have been proved to reach velocities of over 300 miles a second, and subside again with almost equal rapidity. The chromosphere and its quiescent prominences appear to be truly gaseous, consisting of hydrogen, helium, and coronium, while the eruptive prominences always show the presence of metallic vapours, especially of calcium. Prominences increase in size and number in close accordance with the increase of sun-spots. Beyond the red chromosphere and prominences is the marvellous white glory of the corona, which extends to an enormous distance round the sun. Like the prominences of the chromosphere, it is subject to periodical changes in form and size, corresponding to the sun-spot period, but in inverse order, a minimum of sun-spots going with a maximum extension of the corona. At the total eclipse of July 1878, when the sun's surface was almost wholly clear, a pair of enormous equatorial streamers stretched east and west of the sun to a distance of ten millions of miles, and less extensions of the corona occurred at the poles. At the eclipses of 1882 and 1883, on the other hand, when sun-spots were at a maximum, the corona was regularly stellate with no great extensions, but of high brilliancy. This correspondence has been noted at every eclipse, and there is therefore an undoubted connection between the two phenomena.

The light of the corona is believed to be derived from three sources—from incandescent solid or liquid particles thrown out from the sun, from sunlight reflected from these particles, and from gaseous emissions. Its spectrum possesses a green ray, which is peculiar to it, and is supposed to indicate a gas named 'coronium'; in other respects the spectrum is more like that of reflected sunlight. The enormous extensions of the corona into great angular streamers seem to indicate electrical repulsive forces analogous to those which produce the tails of comets.

Connected with the sun's corona is that strange phenomenon, the zodiacal light. This is a delicate nebulosity, which is often seen after sunset in spring and before sunrise in autumn, tapering upwards from the sun's direction along the plane of the ecliptic. Under very favourable conditions it has been traced in the eastern sky in spring to 180° from the sun's position, indicating that it extends beyond the earth's orbit. Long-continued observations from the summit of the Pic du Midi show that this is really the case, and that it lies almost exactly in the plane of the sun's equator. It is therefore held to be produced by the minute particles thrown off the sun, through those coronal wings and streamers which are visible only during solar eclipses.

The careful study of the solar phenomena has very clearly established the fact that none of the sun's envelopes, from the reversing layer to the corona itself, is in any sense an atmosphere. The combination of enormous gravitative force with an amount of heat which turns all the elements into the liquid or gaseous state, leads to consequences which it is difficult for us to follow or comprehend. There is evidently constant internal movement or circulation in the interior of the sun, resulting in the faculæ, the sun-spots, the intensely luminous photosphere, and the chromosphere with its vast flaming coruscations and eruptive protuberances. But it seems impossible that this incessant and violent movement can be kept up without some great and periodical or continuous inrush of fresh materials to renew the heat, keep up the internal circulation, and supply the waste. Perhaps the movement of the sun through space may bring him into contact with sufficiently large masses of matter to continually excite that internal movement without which the exterior surface would rapidly become cool and all planetary life cease. The various solar envelopes are the result of this internal agitation, uprushes, and explosions, while the vast white corona is probably of little more density than comets' tails, probably even of less density, since comets not unfrequently rush through its midst without suffering any loss of velocity. The fact that none of the solar envelopes are visible to us until the light of the photosphere is completely shut off, and that they all vanish the very instant the first gleam of direct sunlight reaches us, is another proof of their extreme tenuity, as is also the sharply defined edge of the sun's disc. The envelopes therefore consist partly of liquid or vaporous matter, in a very finely divided state, driven off by explosions or by electrical forces, and this matter, rapidly cooling, becomes solidified into minutest particles, or even physical molecules. Much of this matter continually falls back on the sun's surface, but a certain quantity of the very finest dust is continually driven away by electrical repulsion, so as to form the corona and the zodiacal light. The vast coronal streamers and the still more extensive ring of the zodiacal light are therefore in all probability due to the same causes, and have a similar physical constitution with the tails of comets.

As the whole of our sunlight must pass through both the reversing layer and the red chromosphere, its colour must be somewhat modified by them. Hence it is believed that, if they were absent, not only would the light and heat of the sun be considerably greater, but its colour would be a purer white, tending towards bluish rather than towards the yellowish tinge it actually possesses.

The Nebular and Meteoritic Hypotheses

As the constitution of the sun, and its agency in producing magnetism and electricity in the matter and orbs around it, afford us our best guide to the constitution of the stars and nebulæ, and to their possible action on each other, and even upon our earth, so the mode of evolution of the sun and solar system, from some pre-existing condition, is likely to help us towards gaining some knowledge of the constitution of the stellar universe and the processes of change going on there.

At the very commencement of the nineteenth century the great mathematician Laplace published his Nebular Theory of the Origin of the Solar System; and although he put it forth merely as a suggestion, and did not support it with any numerical or physical data, or by any mathematical processes, his great reputation, and its apparent probability and simplicity, caused it to be almost universally accepted, and to be extended so as to apply to the evolution of the stellar universe. This theory, very briefly stated, is, that the whole of the matter of the solar system once formed a globular or spheroidal mass of intensely heated gases, extending beyond the orbit of the outermost planet, and having a slow motion of revolution about an axis. As it cooled and contracted, its rate of revolution increased, and this became so great that at successive epochs it threw off rings, which, owing to slight irregularities, broke up, and, gravitating together, formed the planets. The contraction continuing, the sun, as we now see it, was the result.

For about half a century this nebular hypothesis was generally accepted, but during the last thirty years so many objections and difficulties have been suggested, that it has been felt impossible to retain it even as a working hypothesis. At the same time another hypothesis has been put forth which seems more in accordance with the facts of nature as we find them in our own solar system, and which is not open to any of the objections against the nebular theory, even if it introduces a few new ones.

A fundamental objection to Laplace's theory is, that in a gas of such extreme tenuity as the solar nebula must have been, even when it extended only to Saturn or Uranus, it could not possibly have had any cohesion, and therefore could not have given off whole rings at distant intervals, but only small fragments continuously as condensation went on, and these, rapidly cooling, would form solid particles, a kind of meteoric dust, which might aggregate into numerous small planets, or might persist for indefinite periods, like the rings of Saturn or the great ring of the Asteroids.

Another equally vital objection is, that, as the nebula when extending beyond the orbit of Neptune could have had a mean density of only about the two-hundred millionth of our air at sea level, it must have been many hundred times less dense than this at and near its outer surface, and would there be exposed to the cold of stellar space—a cold that would solidify hydrogen. It is thus evident that the gases of all the metallic and other solid elements could not possibly exist as such, but would rapidly, perhaps almost instantaneously, become first liquid and then solid, forming meteoric dust even before contraction had gone far enough to produce such increased rotation as would throw off any portion of the gaseous matter.

Here we have the foundations of the meteoritic hypothesis which is now steadily making its way. It is supported by the fact that we everywhere find proofs of such solid matter in the planetary spaces around us. It falls continually upon the earth. It can be collected on the Arctic and Alpine snows. It occurs everywhere in the deepest abysses of the ocean where there are not sufficient organic deposits to mask it. It constitutes, as has now been demonstrated, the rings of Saturn. Thousands of vast rings of solid particles circulate around the sun, and when our earth crosses any of these rings, and their particles enter our atmosphere with planetary velocity, the friction ignites them and we see falling stars. Comets' tails, the sun's corona, and the zodiacal light are three strange phenomena, which, though wholly insoluble on any theory of gaseous formation, receive their intelligible explanation by means of excessively minute solid particles—microscopic cosmic dust—driven outward by the tremendous electrical repulsions that emanate from the sun.

Having these and other proofs that solid matter, ranging in size, perhaps, from the majestic orbs of Jupiter and Saturn down to the inconceivably minute particles driven millions of miles into space to form a comet's tail, does actually exist everywhere around us, and by collisions between the particles or with planetary atmospheres can produce heat and light and gaseous emanations, we find a basis of fact and observation for the meteoritic hypothesis which Laplace's nebular, and essentially gaseous, theory does not possess.

During the latter half of the nineteenth century several writers suggested this idea of the possible formation of the Solar System, but so far as I am aware, the late R.A. Proctor was the first to discuss it in any detail, and to show that it explained many of the peculiarities in the size and arrangement of the planets and their satellites which the nebular hypothesis did not explain. This he does at some length in the chapter on meteors and comets in his *Other Worlds than Ours*, published in 1870. He assumed, instead of the fire-mist of Laplace, that the space now occupied by the solar system, and for an unknown distance around it, was occupied by vast quantities of solid particles of all the kinds of matter which we now find in the earth, sun, and stars. This matter was dispersed somewhat irregularly, as we see that all the matter of the universe is now distributed; and he further assumed that it was all in motion, as we now know that all the stars and other cosmical masses are, and must be, in motion towards or around some centre.

Under these conditions, wherever the matter was most aggregated, there would be a centre of attraction through gravitation, which would necessarily lead to further aggregation, and the continual impacts of such aggregating matter would produce heat. In course of time, if the supply of cosmic matter was ample (as the result shows that it must have been, whatever theory we adopt), our sun, thus formed, would approximate to its present mass and acquire sufficient heat by collision and gravitation to convert its whole body into the liquid or gaseous condition. While this was going on, subordinate centres of aggregation might form, which would capture a certain proportion of the matter flowing in under the attraction of the central mass, while, owing to the nearly uniform direction and velocity with which the whole system was revolving, each subordinate centre would revolve around the central mass, in somewhat different planes, but all in the same direction.

Mr. Proctor shows the probability that the largest outside aggregation would be at a great distance from the central mass, and this having once been formed, any centres farther away from the sun would be both smaller and very remote, while those inside the first would, as a rule, become smaller as they were nearer the centre. The heated condition of the earth's interior would thus be due, not to the primitive heat of matter in a gaseous state out of which it was formed—a condition physically impossible—but would be acquired in the process of aggregation by the collisions of meteoric masses falling on it, and by its own gravitative force producing continuous condensation and heat.

On this view Jupiter would probably be formed first, and after him at very great distances, Saturn, Uranus, and Neptune; while the inner aggregations would be smaller, as the much greater attractive power of the sun would give them comparatively little opportunity of capturing the meteoric matter that was continuously flowing towards him.

The Meteoritic Nature of the Nebulæ

Having thus reached the conclusion that wherever apparently nebulous matter exists within the limits of the solar system it is not gaseous but consists of solid particles, or, if heated gases are associated with the solid matter they can be accounted for by the heat due to collisions either with other solid particles or with accumulations of gases at a low temperature, as when meteorites enter our atmosphere, it was an easy step to consider whether the cosmic nebulæ and stars may not have had a similar origin.

From this point of view the nebulæ are supposed to be vast aggregations of meteorites or cosmic dust, or of the more persistent gases, revolving with circular or spiral motions, or in irregular streams, and so sparsely scattered that the separate particles of dust may be miles—perhaps hundreds of miles—apart; yet even those nebulæ, only visible by the telescope, may contain as much matter as the whole solar system. From this simple origin, by steps which can be observed in the skies, almost all the forms of suns and systems can be traced by means of the known laws of motion, of heat-production, and of chemical action. The chief English advocate of this view at the present time is Sir Norman Lockyer, who, in numerous papers, and in his works on *The Meteoritic Hypothesis* and *Inorganic Evolution*, has developed it in detail, as the result of many years' continuous research, aided by the contributory work of continental and American astronomers. These views are gradually spreading among astronomers and mathematicians, as will be seen by the very brief outline which will now be given of the explanations they afford of the main groups of phenomena presented by the stellar universe.

Dr. Roberts on Spiral Nebulæ

Dr. Isaac Roberts, who possesses one of the finest telescopes constructed for photographing stars and nebulæ, has given his views on stellar evolution, in *Knowledge* of February 1897, illustrated by four beautiful photographs of spiral nebulæ. These curious forms were at first thought to be rare, but are now found to be really very numerous when details are brought out by the camera. Many of the very large and apparently quite irregular nebulæ, like the Magellanic Clouds, are found to have faint indications of spiral structure. As more than ten thousand nebulæ are now known, and new ones are continually being discovered, it will be a long time before these can all be carefully studied and photographed, but present indications seem to show that a considerable proportion of them will exhibit spiral forms.

Dr. Roberts tells us that all the spiral nebulæ he has photographed are characterised by having a nucleus surrounded by dense nebulosity, most of them being also studded with stars. These stars are always arranged more or less symmetrically, following the curves of the spiral, while outside the visible nebula are other stars arranged in curves strongly suggesting a former greater extension of the nebulous matter. This is so marked a feature that it at once leads to a possible explanation of the numerous slightly curved lines of stars found in every part of the heavens, as being the result of their origin from spiral nebulæ whose material substance has been absorbed by them.

Dr. Roberts proposes several problems in relation to these bodies: Of what materials are spiral nebulæ composed? Whence comes the vortical motion which has produced their forms? The material he finds in those faint clouds of nebulous matter, often of vast extent, that exist in many parts of the sky, and these are so numerous that Sir William Herschel alone recorded the positions of fifty-two such regions, many of which have been confirmed by recent photographs. Dr. Roberts considers these to be either gaseous or with discrete solid particles intermixed. He also enumerates smaller nebulous masses undergoing condensation and segregation into more regular forms; spiral nebulæ in various stages of condensation and of aggregation; elliptic nebulæ; and globular nebulæ. In the last three classes there is clear evidence, on every photograph that has been taken, that condensation into stars or star like forms is now going on.

He adopts Sir Norman Lockyer's view that collisions of meteorites within each swarm or cloud would produce luminous nebulosity; so also would collisions between separate swarms of meteorites produce the conditions required to account for the vortical motions and the peculiar distribution of the nebulosity in the spiral nebulæ. Almost any collision between unequal masses of diffused matter would, in the absence of any massive central body round which they would be forced to revolve, lead to spiral motions. It is to be noted that, although the stars formed in the spiral convolutions of the nebulæ follow those curves, and retain them after the nebulous matter has been all absorbed by them, yet, whenever such a nebula is seen by us edgewise, the convolutions with their enclosed stars will appear as straight lines; and thus not only numbers of star groups arranged in curves, but also those which form almost perfect straight lines, may possibly be traced back to an origin from spiral nebulæ.

Motion being a necessary result of gravitation, we know that every star, planet, comet, or nebula must be in motion through space, and these motions—except in systems physically connected or which have had a common origin—are, apparently, in all directions. How these motions originated and are now regulated we do not know; but there they are, and they furnish the motive power of the collisions, which, when affecting large bodies or masses of diffused matter, lead to the formation of the various kinds of permanent stars; while when smaller masses of matter are concerned those temporary stars are formed which have interested astronomers in all ages. It must be noted that although the motions of the single stars appear to be in straight lines, yet the spaces through which they have been observed to move are so small that they may really be moving in curved orbits around some central body, or the centre of gravity of some aggregation of stars bright and dark, which may itself be comparatively at rest. There may be thousands of such centres around us, and this may sufficiently explain the apparent motions of stars in all directions.

A Suggestion As To the Formation Of Spiral Nebulæ

In a remarkable paper in the Astrophysical Journal (July 1901), Mr. T.C. Chamberlin suggests an origin for the spiral nebulæ, as well as of swarms of meteorites and comets, which seems likely to be a true, although perhaps not the only one.

There is a well-known principle which shows that when two bodies in space, of stellar size, pass within a certain distance of each other, the smaller one will be liable to be torn into fragments by the differential attraction of the larger and denser body. This was originally proved in the case of gaseous and liquid bodies, and the distance within which the smaller one will be disrupted (termed the Roche limit) is calculated on the supposition that the disrupted body is a liquid mass. Mr. Chamberlin shows, however, that a solid body will also be disrupted at a lesser distance dependent on its size and cohesive strength; but, as the size of the two bodies increases, the distance at which disruption will occur increases also, till with very large bodies, such as suns, it becomes almost as large as in the case of liquids or gases.

The disruption occurs from the well-known law of differential gravitation on the two sides of a body leading to tidal deformation in a liquid, and to unequal strain in a solid. When the changes of gravitative force take place slowly, and are also small in amount, the tides in liquids or strains in solids are very small, as in the case of our earth when acted on by the sun and moon, the result is a small tide in the ocean and atmosphere, and no doubt also in the molten interior, to which the comparatively thin crust may partially adjust itself. But if we suppose two dark or luminous suns whose proper motions are in such a direction as to bring them near each other, then, as they approach, each will be deflected towards the other, and will pass round their common centre of gravity with immense velocity, perhaps hundreds of miles in a second. At a considerable distance they will begin to produce tidal elongation towards and away from each other, but when the disruptive limit is nearly reached, the gravitative forces will be increasing so rapidly that even a liquid mass could not adjust its shape with sufficient quickness and the tremendous internal strains would produce the effects of an explosion, tearing the whole mass (of the smaller of the two) into fragments and dust.

But it is also shown that, during the entire process, the two elongated portions of the originally spherical mass would be so acted upon by gravity as to produce increasing rotation, which as the crisis approached would extend the elongation, and aid in the explosive result. This rapid rotation of the elongated mass would, when the disruption occurred, necessarily give to the fragments a whirling or spiral motion, and thus initiate a spiral nebula of a size and character dependent on the size and constitution of the two masses, and on the amount of the explosive forces set up by their approach.

There is one very suggestive phenomenon which seems to prove that this *is* one of the modes of formation of spiral nebulæ. When the explosive disruption occurs the two protuberances or elongations of the body will fly apart, and having also a rapid rotatory movement, the resulting spiral will necessarily be a double one. Now, it is the fact that almost all the well-developed spiral nebulæ have two such arms opposite to each other, as beautifully shown in M. 100 Comæ, M. 51 Canum, and others photographed by Dr. I. Roberts. It does not seem likely that any other origin of these nebulæ should give rise to a double rather than to a single spiral.

The Evolution of Double Stars

The advance in knowledge of double and multiple stars has been wonderfully rapid, numerous observers having devoted themselves to this special branch. Many thousands were discovered during the first half of the nineteenth century, and as telescopic power increased new ones continued to flow in by hundreds and thousands, and there has been recently published by the Yerkes Observatory a catalogue of 1290 such stars, discovered between 1871 and 1899 by one observer, Mr. S.W. Burnham. All these have been found by the use of the telescope, but during the last quarter of a century the spectroscope has opened up a new world of double stars of enormous extent and the highest interest.

The telescopic binaries which have been observed for a sufficient time to determine their orbits, range from periods of about eleven years as a minimum up to hundreds and even more than a thousand years. But the spectroscope reveals the fact that the many thousands of telescopic binaries form only a very small part of the binary systems in existence. The overwhelming importance of this discovery is, that it carries the times of revolution from the minimum of the telescopic doubles downward in unbroken series through periods of a few years, to those reckoned by months, by days, and even by hours. And with this reduction of period there necessarily follows a corresponding reduction of distance, so that sometimes the two stars must be in contact, and thus the actual birth or origin of a double star has been observed to occur, even though not actually seen. This mode of origin was indeed anticipated by Dr. Lee of Chicago in 1892, and it has been confirmed by observation in the short space of ten years.

In a remarkable communication to *Nature* (September 12th, 1901) Mr. Alexander W. Roberts of Lovedale, South Africa, gives some of the main results of this branch of inquiry.

Of course all the variable stars are to be found among the spectroscopic binaries. They consist of that portion of the class in which the plane of the orbit is directed towards us, so that during their revolution one of the pair either wholly or partially eclipses the other. In some of these cases there are irregularities, such as double maxima and minima of unequal lengths, which may be due to triple systems or to other causes not yet explained, but as they all have short periods and always appear as one star in the most powerful telescopes, they form a special division of the spectroscopic binary systems.

There are known at present twenty-two variables of the Algol type, that is, stars having each a dark companion very close to it which obscures it either wholly or partially during every revolution. In these cases the density of the systems can be approximately determined, and they are found to be, on the average, only one-fifth that of water, or one-eighth that of our sun. But as many of them are as large as our sun, or even considerably larger, it is evident that they must be wholly gaseous, and, even if very hot, of a less complex constitution than our luminary. Mr. A.W. Roberts tells us that five out of these twenty-two variables revolve *in absolute contact* forming systems of the shape of a dumb-bell. The periods vary from twelve days to less than nine hours; and, starting from these, we now have a continuous series of lengthening periods up to the twin stars of Castor which require more than a thousand years to complete their revolution.

During his observations of the above five stars, Mr. Roberts states that one, X Carinæ, was found to have parted company, so that instead of being actually united to its companion the two are now at a distance apart equal to one-tenth of their diameters, and he may thus be said to have been almost a witness of the birth of a stellar system.

A year later we find the record (in *Knowledge*, October 1902) of Professor Campbell's researches at the Lick Observatory. He states that, out of 350 stars observed spectroscopically, one in eight is a spectroscopic binary; and so impressed is he with their abundance that, as accuracy of measurement increases, he believes that *the star that is not a spectroscopic binary will prove to be the rare exception*! Professor G. Darwin had already shown that the 'dumb-bell' was a figure of equilibrium in a rotating mass of fluid; and we now find proofs that such figures exist, and that they form the starting-point for the enormous and ever-increasing quantities of spectroscopic binary star-systems that are now known. The origin of these binary stars is also of especial interest as giving support to Professor Darwin's well-known explanation of the origin of the moon by disruption from the earth, owing to the very rapid rotation of the parent planet. It now appears that suns often subdivide in the same manner, but, owing perhaps to their intensely heated gaseous state they seem usually to form nearly equal globes. The evolution of this special form of star-system is therefore now an observed fact; though it by no means follows that all double stars have had the same mode of origin.

Clusters of Stars and Variables

The clusters of stars, which are tolerably abundant in the heavens and offer so many strange and beautiful forms to the telescopist, are yet among the most puzzling phenomena the philosophic astronomer has to deal with.

Many of these clusters which are not very crowded and of irregular forms, strongly suggest an origin from the equally irregular and fantastic forms of nebulæ by a process of aggregation like that which Dr. Roberts describes as developing within the spiral nebulæ. But the dense globular clusters which form such beautiful telescopic objects, and in some of which more than six thousand stars have been counted besides considerable numbers so crowded in the centre as to be uncountable, are more difficult to explain. One of the problems suggested by these clusters is as to their stability. Professor Simon Newcomb remarks on this point as follows: 'Where thousands of stars are condensed into a space so small, what prevents them from all falling together into one confused mass? Are they really doing so, and will they ultimately form a single body? These are questions which can be satisfactorily answered only by centuries of observation; they must therefore be left to the astronomers of the future.'

There are, however, some remarkable features in these clusters which afford possible indications of their origin and essential constitution. When closely examined most of them are seen to be less regular than they at first appear. Vacant spaces can be noted in them; even rifts of definite forms. In some there is a radiated structure; in others there are curved appendages; while some have fainter centres. These features are so exactly like what are found, in a more pronounced form, in the larger nebulæ, that we can hardly help thinking that in these clusters we have the result of the condensation of very large nebulæ, which have first aggregated towards numerous centres, while these agglomerations have been slowly drawn towards the common centre of gravity of the whole mass. It is suggestive of this origin that while the smaller telescopic nebulæ are far removed from the Milky Way, the larger ones are most abundant near its borders; while the star-clusters are excessively abundant on and near the Milky Way, but very scarce elsewhere, except in or near vast nebulæ like the Magellanic Clouds. We thus see that the two phenomena may be complementary to each other, the condensation of nebulæ having gone on most rapidly where material was most abundant, resulting in numerous star-clusters where there are now few nebulæ.

There is one striking feature of the globular clusters which calls for notice; the presence in some of them of enormous quantities of variable stars, while in others few or none can be found. The Harvard Observatory has for several years devoted much time to this class of observations, and the results are given in Professor Newcomb's recent volume on 'The Stars.' It appears that twenty-three clusters have been observed spectroscopically, the number of stars examined in each cluster varying from 145 up to 3000, the total number of stars thus minutely tested being 19,050. Out of this total number 509 were found to be variable; but the curious fact is, the extreme divergence in the proportion of variables to the whole number examined in the several clusters. In two clusters, though 1279 stars were examined, not a single variable was found. In three others the proportion was from one in 1050 to one in 500. Five more ranged up to one in 100, and the remainder showed from that proportion up to one in seven, 900 stars being examined in the last mentioned cluster of which 132 were variable!

When we consider that variable stars form only a portion, and necessarily a very small proportion, of binary systems of stars, it follows that in all the clusters which show a large proportion of variables, a very much larger proportion—in some cases perhaps all, must be double or multiple stars revolving round each other. With this remarkable evidence, in addition to that adduced for the prevalence of double stars and variables among the stars in general, we can understand Professor Newcomb adding his testimony to that of Professor Campbell already quoted, that 'it is probable that among the stars in general, single stars are the exception rather than the rule. If such be the case, the rule should hold yet more strongly among the stars of a condensed cluster.'

The Evolution of the Stars

So long as astronomers were limited to the use of the telescope only, or even the still greater powers of the photographic plate, nothing could be learnt of the actual constitution of the stars or of the process of their evolution. Their apparent magnitudes, their movements, and even the distances of a few could be determined; while the diversity of their colours offered the only clue (a very imperfect one) even to their temperature. But the discovery of spectrum analysis has furnished the means of obtaining some definite knowledge of the physics and chemistry of the stars, and has thus established a new branch of science—Astrophysics—which has already attained large proportions, and which furnishes the materials for a periodical and some important volumes. This branch of the subject is very complex, and as it is not directly connected with our present inquiry, it is only referred to again in order to introduce such of its results as bear upon the question of the classification and evolution of the stars.

By a long series of laboratory experiments it has been shown that numerous changes occur in the spectra of the elements when subjected to different temperatures, ranging upwards to the highest attainable by means of a battery producing an electric spark several feet long. These changes are not in the relative position of the bands or dark lines, but in their number, breadth, and intensity. Other changes are due to the density of the medium in which the elements are heated, and to their chemical condition as to purity; and from these various modifications and their comparison with the solar spectrum and those of its appendages, it has become possible to determine, from the spectrum of a star, not only its temperature as compared with that of the electric spark and of the sun, but also its place in a developmental series.

The first general result obtained by this research is, that the bluish white or pure white stars, having a spectrum extending far towards the violet end, and which exhibits the coloured bands of gases only, usually hydrogen and helium, are the hottest. Next come those with a shorter spectrum not extending so far towards the violet end, and whose light is therefore more yellow in tint. To this group our sun belongs; and they are all characterised like it by dark lines due to absorption, and by the presence of metals, especially iron, in a gaseous state. The third group have the shortest spectra and are of a red colour, while their spectra contain lines denoting the presence of carbon. These three groups are often spoken of as 'gaseous stars,' 'metallic stars,' and 'carbon stars.' Other astronomers call the first group 'Sirian stars,' because Sirius, though not the hottest, is a characteristic type; the second being termed 'solar stars'; others again speak of them as stars of Class I., Class II., etc., according to the system of classification they have adopted. It was soon perceived, however, that neither the colour nor the temperature of stars gave much information as to their nature and state of development, because, unless we supposed the stars to begin their lives already intensely hot (and all the evidence is against this), there must be a period during which heat increases, then one of maximum heat, followed by one of cooling and final loss of light altogether. The meteoritic theory of the origin of all luminous bodies in the heavens, now very widely adopted, has been used, as we have seen, to explain the development of stars from nebulæ, and its chief exponent in this country, Sir Norman Lockyer, has propounded a complete scheme of stellar evolution and decay which may be here briefly outlined:

Beginning with nebulæ, we pass on to stars having banded or fluted spectra, indicating comparatively low temperatures and showing bands or lines of iron, manganese, calcium, and other metals. They are more or less red in colour, Antares in the Scorpion being one of the most brilliant red stars known. These stars are supposed to be in the process of aggregation, to be continually increasing in size and heat, and thus to be subject to great disturbances. Alpha Cygni has a similar spectrum but with more hydrogen, and is much hotter. The increase of heat goes on through Rigel and Beta Crucis, in which we find mainly hydrogen, helium, oxygen, nitrogen, and also carbon, but only faint traces of metals. Reaching the hottest of all—Epsilon Orionis and two stars in Argo—hydrogen is predominant, with traces of a few metals and carbon. The cooling series is indicated by thicker lines of hydrogen and thinner lines of the metallic elements, through Sirius, to Arcturus and our sun, thence to 19 Piscium, which shows chiefly flutings of carbon, with a few faint metallic lines. The process of further cooling brings us to the dark stars.

We have here a complete scheme of evolution, carrying us from those ill-defined but enormously diffused masses of gas and cosmic dust we know as nebulæ, through planetary nebulæ, nebulous stars, variable and double-stars, to red and white stars and on to those exhibiting the most intense blue-white lustre. We must remember, however, that the most brilliant of these stars, showing a gaseous spectrum and forming the culminating point of the ascending series, are not necessarily hotter than, or even so hot as, some of those far down on the descending scale; since it is one of the apparent paradoxes of physics that a body may become hotter during the very process of contraction through loss of heat. The reason is that by cooling it contracts and thus becomes denser, that a portion of its mass falls towards its centre, and in doing so produces an amount of heat which, though absolutely less than the heat lost in cooling, will under certain conditions cause the reduced surface to become hotter. The essential point is, that the body in question must be wholly gaseous, allowing of free circulation from surface to centre. The law, as given by Professor S. Newcomb, is as follows:—

'*When a spherical mass of incandescent gas contracts through the loss of its heat by radiation into space, its temperature continually becomes higher as long as the gaseous condition is retained.*'

To put it in another way: if the compression was caused by external force and no heat was lost, the globe would get hotter by a calculable amount for each unit of contraction. But the heat lost in causing a similar amount of contraction is so little more than the increase of heat produced by contraction, that the slightly diminished total heat in a smaller bulk causes the temperature of the mass to increase.

But if, as there is reason to believe, the various types of stars differ also in chemical constitution, some consisting mainly of the more permanent gases, while in others the various metallic and non-metallic elements are present in very different proportions, there should really be a classification by constitution as well as by temperature, and the course of evolution of the differently constituted groups may be to some extent dissimilar.

With this limitation the process of evolution and decay of sun through a cycle of increasing and decreasing temperature, as suggested by Sir Norman Lockyer, is clear and suggestive. During the ascending series the star is growing both in mass and heat, by the continual accretion of meteoritic matter either drawn to it by gravitation or falling towards it through the proper motions of independent masses. This goes on till all the matter for some distance around the star has been utilised, and a maximum of size, heat, and brilliancy attained. Then the loss of heat by radiation is no longer compensated by the influx of fresh matter, and a slow contraction occurs accompanied by a slightly increased temperature. But owing to the more stable conditions continuous envelopes of metals in the gaseous state are formed, which check the loss of heat and reduce the brilliancy of colour; whence it follows that bodies like our sun may be really hotter than the most brilliant white stars, though not giving out quite so much heat. The loss of heat is therefore reduced; and this may serve to account for the undoubted fact that during the enormous epochs of geological time there has been very little diminution in the amount of heat we have received from the sun.

On the general question of the meteoritic hypothesis one of our first mathematicians, Professor George Darwin, has thus expressed his views: 'The conception of the growth of the planetary bodies by the aggregation of meteorites is a good one, and perhaps seems more probable than the hypothesis that the whole solar system was gaseous.' I may add, that one of the chief objections made to it, that meteorites are too complex to be supposed to be the primitive matter out of which suns and worlds have been made, does not seem to me valid. The primitive matter, whatever it was, may have been used up again and again, and if collisions of large solid globes ever occur—and it is assumed by most astronomers that they must sometimes occur—then meteoric particles of all sizes would be produced which might exhibit any complexity of mineral constitution. The material universe has probably been in existence long enough for all the primitive elements to have been again and again combined into the minerals found upon the earth and many others. It cannot be too often repeated that no explanation—no theory—can ever take us to the beginning of things, but only one or two steps at a time into the dim past, which may enable us to comprehend, however imperfectly, the processes by which the world, or the universe, as it is, has been developed out of some earlier and simpler condition.

CHAPTER VII

ARE THE STARS INFINITE IN NUMBER?

Most of the critics of my first short discussion of this subject laid great stress upon the impossibility of proving that the universe, a part of which we see, is not infinite; and a well-known astronomer declared that unless it can be demonstrated that our universe is finite the entire argument founded upon our position within it fall to the ground. I had laid myself open to this objection by rather incautiously admitting that if the preponderance of evidence pointed in this direction any inquiry as to our place in the universe would be useless, because as regards infinity there can be no difference of position. But this statement is by no means exact, and even in an infinite universe of matter containing an infinite number of stars, such as those we see, there might well be such infinite diversities of distribution and arrangement as would give to certain positions all the advantages which I submit we actually possess. Supposing, for example, that beyond the vast ring of the Milky Way the stars rapidly decrease in number in all directions for a distance of a hundred or a thousand times the diameter of that ring, and that then for an equal distance they slowly increase again and become aggregated into systems or universes totally distinct from ours in form and structure, and so remote that they can influence us in no way whatever. Then, I maintain, our position within our own stellar universe might have exactly the same importance, and be equally suggestive, as if ours were the only material universe in existence—as if the apparent diminution in the number of stars (which is an observed fact) indicated a continuous diminution, leading at some unknown distance to entire absence of luminous— that is, of active, energy-emitting aggregations of matter. As to whether there are such other material universes or not I offer no opinion, and have no belief one way or the other. I consider all speculations as to what may or may not exist in infinite space to be utterly valueless. I have limited my inquiries strictly to the evidence accumulated by modern astronomers, and to direct inferences and logical deductions from that evidence. Yet, to my great surprise, my chief critic declares that 'Dr. Wallace's underlying error is, indeed, that he has reasoned from the area which we can embrace with our limited perceptions to the infinite beyond our mental or intellectual grasp.' I have distinctly *not* done this, but many astronomers have done so. The late Richard Proctor not only continually discussed the question of infinite matter as well as infinite space, but also argued, from the supposed attributes of the Deity, for the necessity of holding this material universe to be infinite, and the last chapter of his *Other Worlds than Ours* is mainly devoted to such speculations. In a later work, *Our Place among Infinities*, he says that 'the teachings of science bring us into the presence of the unquestionable infinities of time and of space, and the presumable infinities of matter and of operation—hence therefore into the presence of infinity of energy. But science teaches us nothing about these infinities as such. They remain none the less inconceivable, however clearly we may be taught to recognise their reality.' All this is very reasonable, and the last sentence is particularly important. Nevertheless, many writers allow their reasonings from facts to be influenced by these ideas of infinity. In Proctor's posthumous work, *Old and New Astronomy*, the late Mr. Ranyard, who edited it, writes: 'If we reject as abhorrent to our minds the supposition that the universe is not infinite, we are thrown back on one of two alternatives—either the ether which transmits the light of the stars to us is not perfectly elastic, or a large proportion of the light of the stars is obliterated by dark bodies.' Here we have a well-informed astronomer allowing his abhorrence of the idea of a finite universe to affect his reasoning on the actual phenomena we can observe—doing in fact exactly what my critic erroneously accuses me of doing. But setting aside all ideas and prepossessions of the kind here indicated, let us see what are the actual facts revealed by the best instruments of modern astronomy, and what are the natural and logical inferences from those facts.

Are the Stars Infinite in Number?

The views of those astronomers who have paid attention to this subject are, on the whole, in favour of the view that the stellar universe is limited in extent and the stars therefore limited in number. A few quotations will best exhibit their opinions on this question, with some of the facts and observations on which they are founded.

Miss A.M. Clerke, in her admirable volume, *The System of the Stars*, says: 'The sidereal world presents us, to all appearance, with a finite system.... The probability amounts almost to certainty that star-strewn space is of measurable dimensions. For from innumerable stars a limitless sum-total of radiations should be derived, by which darkness would be banished from our skies; and the "intense inane," glowing with the mingled beams of suns individually indistinguishable, would bewilder our feeble senses with its monotonous splendour.... Unless, that is to say, light suffer some degree of enfeeblement in space.... But there is not a particle of evidence that any such toll is exacted; contrary indications are strong; and the assertion that its payment is inevitable depends upon analogies which may be wholly visionary. We are then, for the present, entitled to disregard the problematical effect of a more than dubious cause.'

Professor Simon Newcomb, one of the first of American mathematicians and astronomers, arrives at a similar conclusion in his most recent volume, *The Stars* (1902). He says, in his conclusions at the end of the work: 'That collection of stars which we call the universe is limited in extent. The smallest stars that we see with the most powerful telescopes are not, for the most part, more distant than those a grade brighter, but are mostly stars of less luminosity situate in the same regions' (p. 319). And on page 229 of the same work he gives reasons for this conclusion, as follows: 'There is a law of optics which throws some light on the question. Suppose the stars to be scattered through infinite space so that every great portion of space is, in the general average, equally rich in stars. Then at some great distance we describe a sphere having its centre in our sun. Outside this sphere describe another one of a greater radius, and beyond this other spheres at equal distances apart indefinitely. Thus we shall have an endless succession of spherical shells, each of the same thickness. The volume of each of these shells will be nearly proportional to the squares of the diameters of the spheres which bound it. Hence each of the regions will contain a number of stars increasing as the square of the radius of the region. Since the amount of light we receive from each star is as the inverse square of its distance, it follows that the sum total of the light received from each of these spherical shells will be equal. Thus as we add sphere after sphere we add equal amounts of light without limit. The result would be that if the system of stars extended out indefinitely the whole heavens would be filled with a blaze of light as bright as the sun.'

But the whole light given us by the stars is variously estimated at from one-fortieth to one-twentieth or, as an extreme limit, to one-tenth of moonlight, while the sun gives as much light as 300,000 full moons, so that starlight is only equivalent at a fair estimate to the six-millionth part of sunlight. Keeping this in mind, the possible causes of the extinction of almost the whole of the light of the stars (if they are infinite in number and distributed, on the average, as thickly beyond the Milky Way as they are up to its outer boundary) are absurdly inadequate. These causes are (1) the loss of light in passing through the ether, and (2) the stoppage of light by dark stars or diffused meteoritic dust. As to the first, it is generally admitted that there is not a particle of evidence of its existence. There is, however, some distinct evidence that, if it exists, it is so very small in amount that it would not produce a perceptible effect for any distances less remote than hundreds or perhaps thousands of times as far as the farthest limits of the Milky Way are from us. This is indicated by the fact that the brightest stars are *not* always, or even generally, the nearest to us, as is shown both by their small proper motions and the absence of measurable parallax. Mr. Gore states that out of twenty-five stars, with proper motions of more than two seconds annually, only two are above the third magnitude. Many first magnitude stars, including Canopus, the second brightest star in the heavens, are so remote that no parallax can be found, notwithstanding repeated efforts. They must therefore be much farther off than many small and telescopic stars, and perhaps as far as the Milky Way, in which so many brilliant stars are found; whereas if any considerable amount of light were lost in passing that distance we should find but few stars of the first two or three magnitudes that were very remote from us. Of the twenty-three stars of the first magnitude, only ten have been found to have parallaxes of more than one-twentieth of a second, while five range from that small amount down to one or two hundredths of a second, and there are two with no ascertainable parallax. Again, there are 309 stars brighter than magnitude 3.5, yet only thirty-one of these have proper motions of more than 100" a century, and of these only eighteen have parallaxes of more than one-twentieth of a second. These figures are from tables given in Professor Newcomb's book, and they have very great significance, since they indicate that the brightest stars are *not* the nearest to us. More than this, they show that out of the seventy-two stars whose distance has been measured with some approach to certainty, only twenty-three (having a parallax of more than one-fiftieth of a second) are of greater magnitudes than 3.5, while no less than forty-nine are smaller stars down to the eighth or ninth magnitude, and these are on the average much nearer to us than the brighter stars!

Taking the whole of the stars whose parallaxes are given by Professor Newcomb, we find that the average parallax of the thirty-one bright stars (from 3.5 magnitude up to Sirius) is 0.11 seconds; while that of the forty-one stars below 3.5 magnitude down to about 9.5, is 0.21 seconds, showing that they are, on the average, only half as far from us as the brighter stars. The same conclusion was reached by Mr. Thomas Lewis of the Greenwich Observatory in 1895, namely, that the stars from 2.70 magnitude down to about 8.40 magnitude have, on the average, double the parallaxes of the brighter stars. This very curious and unexpected fact, however it may be accounted for, is directly opposed to the idea of there being any loss of light by the more distant as compared with the nearer stars; for if there should be such a loss it would render the above phenomenon still more difficult of explanation, because it would tend to exaggerate it. The bright stars being on the whole farther away from us than the less bright down to the eighth and ninth magnitudes, it follows, if there is any loss of light, that the bright stars are really brighter than they appear to us, because, owing to their enormous distance some of their light has been lost before it reached us. Of course it may be said that this does not *demonstrate* that no light is lost in passing through space; but, on the other hand, it is exactly the opposite of what we should expect if the more distant stars were perceptibly dimmed by this cause, and it may be considered to prove that if there is any loss it is exceedingly small, and will not affect the question of the limits of our stellar system, which is all that we are dealing with.

This remarkable fact of the enormous remoteness of the majority of the brighter stars is equally effective as an argument against the loss of light by dark stars or cosmic dust, because, if the light is not appreciably diminished for stars which have less than the fiftieth of a second of parallax, it cannot greatly interfere with our estimates of the limits of our universe.

Both Mr. E.W. Maunder of the Greenwich Observatory and Professor W.W. Turner of Oxford lay great stress on these dark bodies, and the former quotes Sir Robert Ball as saying, 'the dark stars are incomparably more numerous than those that we can see ... and to attempt to number the stars of our universe by those whose transitory brightness we can perceive would be like estimating the number of horseshoes in England by those which are red-hot.' But the proportion of dark stars (or nebulæ) to bright ones cannot be determined *a priori*, since it must depend upon the causes that heat the stars, and how frequently those causes come into action as compared with the life of a bright star. We do know, both from the stability of the light of the stars during the historic period and much more precisely by the enormous epochs during which our sun has supported life upon this earth—yet which must have been 'incomparably' less than its whole existence as a light-giver—that the life of most stars must be counted by hundreds or perhaps by thousands of millions of years. But we have no knowledge whatever of the rate at which true stars are born. The so-called 'new stars' which occasionally appear evidently belong to a different category. They blaze out suddenly and almost as suddenly fade away into obscurity or total invisibility. But the true stars probably go through their stages of origin, growth, maturity, and decay, with extreme slowness, so that it is not as yet possible for us to determine by observation when they are born or when they die. In this respect they correspond to species in the organic world. They would probably first be known to us as stars or minute nebulæ: at the extreme limit of telescopic vision or of photographic sensitiveness, and the growth of their luminosity might be so gradual as to require hundreds, perhaps thousands of years to be distinctly recognisable. Hence the argument derived from the fact that we have never witnessed the birth of a true permanent star, and that, therefore, such occurrences are very rare, is valueless. New stars may arise every year or every day without our recognising them; and if this is the case, the reservoir of dark bodies, whether in the form of large masses or of clouds of cosmic dust, so far from being incomparably greater than the whole of the visible stars and nebulæ, may quite possibly be only equal to it, or at most a few times greater; and in that case, considering the enormous distances that separate the stars (or star-systems) from each other, they would have no appreciable effect in shutting out from our view any considerable proportion of the luminous bodies constituting our stellar universe. It follows, that Professor Newcomb's argument as to the very small total light given by the stars has not been even weakened by any of the facts or arguments adduced against it.

Mr. W.H.S. Monck, in a letter to *Knowledge* (May 1903), puts the case very strongly so as to support my view. He says:—'The highest estimate that I have seen of the total light of the full moon is $^1/_{300000}$ of that of the sun. Suppose that the dark bodies were a hundred and fifty thousand times as numerous as the bright ones. Then the whole sky ought to be as bright as the illuminated portion of the moon. Every one knows that this is not so. But it is said that the stars, though infinite, may only extend to infinity in particular directions, *e.g.* in that of the Galaxy. Be it so. Where, in the very brightest portion of the Galaxy, will we find a part equal in angular magnitude to the moon which affords us the same quantity of light? In the very brightest spot, the light probably does not amount to one hundredth part that of the full moon.' It follows that, even if dark stars were fifteen million times as numerous as the bright ones, Professor Newcomb's argument would still apply against an infinite universe of stars of the same average density as the portion we see.

Telescopic Evidence as to the Limits of the Star System

Throughout the earlier portion of the nineteenth century every increase of power and of light-giving qualities of telescopes added so greatly to the number of the stars which became visible, that it was generally assumed that this increase would go on indefinitely, and that the stars were really infinite in number and could not be exhausted. But of late years it has been found that the increase in the number of stars visible in the larger telescopes was not so great as might be expected, while in many parts of the heavens a longer exposure of the photographic plate adds comparatively little to the number of stars obtained by a shorter exposure with the same instrument.

Mr. J.E. Gore's testimony on this point is very clear. He says:—'Those who do not give the subject sufficient consideration, seem to think that the number of the stars is practically infinite, or at least, that the number is so great that it cannot be estimated. But this idea is totally incorrect, and due to complete ignorance of telescopic revelations. It is certainly true that, to a certain extent, the larger the telescope used in the examination of the heavens, the more the number of the stars seems to increase; but we now know that there is a limit to this increase of telescopic vision. And the evidence clearly shows that we are rapidly approaching this limit. Although the number of stars visible in the Pleiades rapidly increases at first with increase in the size of the telescope used, and although photography has still further increased the number of stars in this remarkable cluster, it has recently been found that an increased length of exposure—beyond three hours—adds very few stars to the number visible on the photograph taken at the Paris Observatory in 1885, on which over two thousand stars can be counted. Even with this great number on so small an area of the heavens, comparatively large vacant places are visible between the stars, and a glance at the original photograph is sufficient to show that there would be ample room for many times the number actually visible. I find that if the whole heavens were as rich in stars as the Pleiades, there would be only thirty-three millions in both hemispheres.'

Again, referring to the fact that Celoria, with a telescope showing stars down to the eleventh magnitude, could see almost exactly the same number of stars near the north pole of the Galaxy as Sir William Herschel found with his much larger and more powerful telescope, he remarks: 'Their absence, therefore, seems certain proof that very faint stars do *not* exist in that direction, and that here, at least, the sidereal universe is limited in extent.'

Sir John Herschel notes the same phenomena, stating that even in the Milky Way there are found 'spaces absolutely dark *and completely void of any star*, even of the smallest telescopic magnitude'; while in other parts 'extremely minute stars, though never altogether wanting, occur in numbers so moderate as to lead us irresistibly to the conclusion that in these regions we see *fairly through* the starry stratum, since it is impossible otherwise (supposing their light not intercepted) that the numbers of the smaller magnitudes should not go on continually increasing ad infinitum. In such cases, moreover, the ground of the heavens, as seen between the stars, is for the most part perfectly dark, which again would not be the case if innumerable multitudes of stars, too minute to be individually discernible, existed beyond.' And again he sums up as follows:—'Throughout by far the larger portion of the extent of the Milky Way in both hemispheres, the general blackness of the ground of the heavens on which its stars are projected, and the absence of that innumerable multitude and excessive crowding of the smallest visible magnitudes, and of glare produced by the aggregate light of multitudes too small to affect the eye singly, which the contrary supposition would appear to necessitate, must, we think, be considered unequivocal indications that its dimensions *in directions where these conditions obtain*, are not only not infinite, but that the space-penetrating power of our telescopes suffices fairly to pierce through and beyond it.'

This expression of opinion by the astronomer who, probably beyond any now living, was the most competent authority on this question, to which he devoted a long life of observation and study extending over the whole heavens, cannot be lightly set aside by the opinions or conjectures of those who seem to assume that we must believe in an infinity of stars if the contrary cannot be absolutely proved. But as not a particle of evidence can be adduced to prove infinity, and as all the facts and indications point, as here shown, in a directly opposite direction, we must, if we are to trust to evidence at all in this matter, arrive at the conclusion that the universe of stars is limited in extent.

Dr. Isaac Roberts gives similar evidence as regards the use of photographic plates. He writes:—'Eleven years ago photographs of the Great Nebula in *Andromeda* were taken with the 20-inch reflector, and exposures of the plates during intervals up to four hours; and upon some of them were depicted stars to the faintness of 17th to 18th magnitude, and nebulosity to an equal degree of faintness. The films of the plates obtainable in those days were less sensitive than those which have been available during the past five years, and during this period photographs of the nebula with exposures up to four hours have been taken with the 20-inch reflector. No extensions of the nebulosity, however, nor increase in the number of the stars can be seen on the later rapid plates than were depicted upon the earlier slower ones, though the star-images and the nebulosity have greater density on the later plates.'

Exactly similar facts are recorded in the cases of the

Great Nebula in *Orion*, and the group of the Pleiades. In the case of the Milky Way in *Cygnus* photographs have been taken with the same instrument, but with exposures varying from one hour to two hours and a half, but no fainter stars could be found on one than on the other; and this fact has been confirmed by similar photographs of other areas in the sky.

The Law of Diminishing Numbers of Stars

We will now consider another kind of evidence equally weighty with the two already adduced. This is what may be termed the law of diminishing numbers beyond a certain magnitude, as observed by larger and larger telescopes.

For some years past star-magnitudes have been determined very accurately by means of careful photometric comparisons. Down to the sixth magnitude stars are visible to the naked eye, and are hence termed lucid stars. All fainter stars are telescopic, and continuing the magnitudes in a series in which the difference in luminosity between each successive magnitude is equal, the seventeenth magnitude is reached and indicates the range of visibility in the largest telescopes now in existence. By the scale now used a star of any magnitude gives nearly two and a half times as much light as one of the next lower magnitude, and for accurate comparison the apparent brightness of each star is given to the tenth of a magnitude which can easily be observed. Of course, owing to differences in the colour of stars, these determinations cannot be made with perfect accuracy, but no important error is due to this cause. According to this scale a sixth magnitude star gives about one-hundredth part of the light of an average first magnitude star. Sirius is so exceptionally bright that it gives nine times as much light as a standard or average first magnitude star.

Now it is found that from the first to the sixth magnitude the stars increase in number at the rate of about three and a half times those of the preceding magnitudes. The total number of stars down to the sixth magnitude is given by Professor Newcomb as 7647. For higher magnitudes the numbers are so great that precision and uniformity are more difficult of attainment; yet there is a wonderful continuance of the same law of increase down to the tenth magnitude, which is estimated to include 2,311,000 stars, thus conforming very nearly with the ratio of 3.5 as determined by the lucid stars.

But when we pass beyond the tenth magnitude to those vast numbers of faint stars only to be seen in the best or the largest telescopes, there appears to be a sudden change in the ratio of increased numbers per magnitude. The numbers of these stars are so great that it is impossible to count the whole as with the higher magnitude stars, but numerous counts have been made by many astronomers in small measured areas in different parts of the heavens, so that a fair average has been obtained, and it is possible to make a near approximation to the total number visible down to the seventeenth magnitude. The estimate of these by astronomers who have made a special study of this subject is, that the total number of visible stars does not exceed one hundred millions.

But if we take the number of stars down to the ninth magnitude, which are known with considerable accuracy, and find the numbers in each succeeding magnitude down to the seventeenth, according to the same ratio of increase which has been found to correspond very nearly in the case of the higher magnitudes, Mr. J.E. Gore finds that the total number should be about 1400 millions. Of course neither of these estimates makes any pretence to exact accuracy, but they are founded on all the facts at present available, and are generally accepted by astronomers as being the nearest approach that can be made to the true numbers. The discrepancy is, however, so enormous that probably no careful observer of the heavens with very large telescopes doubts that there is a very real and very rapid diminution in the numbers of the fainter as compared with the brighter stars.

There is, however, yet one more indication of the decreasing numbers of the faint telescopic stars, which is almost conclusive on this question, and, so far as I am aware, has not yet been used in this relation. I will therefore briefly state it.

The Light Ratio as Indicating the Number of Faint Stars

Professor Newcomb points out a remarkable result depending on the fact that, while the average light of successively lower magnitudes diminishes in a ratio of 2.5, their numbers increase at nearly a ratio of 3.5. From this it follows that, so long as this law of increase continues, the total of starlight goes on increasing by about forty per cent. for each successive magnitude, and he gives the following table to illustrate it:—

Mag.		Total Light	
Mag.	1		=1
"	2	"	= 1.4
"	3	"	= 2.0
"	4	"	= 2.8
"	5	"	= 4.0
"	6	"	= 5.7
"	7	"	= 8.0
"	8	"	= 11.3
"	9	"	= 16.0
"	10	"	= 22.6

Total light to Mag. 10 = 74.8

Thus the total amount of the light given by all stars down to the tenth magnitude is seventy-four times as great as that from the few first magnitude stars. We also see that the light given by the stars of any magnitude is twice as much as that of the stars two magnitudes higher in the scale, so that we can easily calculate what additional light we ought to receive from each additional magnitude if they continue to increase in numbers below the tenth as they do above that magnitude. Now it has been calculated as the result of careful observations, that the total light given by stars down to nine and a half magnitude is one-eightieth of full moonlight, though some make it much more. But if we continue the table of light-ratios from this low starting-point down to magnitude seventeen and a half, we shall find, if the numbers of the stars go on increasing at the same rate as before, that the light of all combined should be at least seven times as great as moonlight; whereas the photometric measurements make it actually about one-twentieth. And as the calculation from light-ratios only includes stars just visible in the largest telescopes, and does not include all those proved to exist by photography, we have in this case a demonstration that the numbers of the stars below the tenth and down to the seventeenth magnitude diminish rapidly.

We must remember that the minuter telescopic stars preponderate enormously in and near the Milky Way. At a distance from it they diminish rapidly, till near its poles they are almost entirely absent. This is shown by the fact (already referred to at p. 146) that Professor Celoria of Milan, with a telescope of less than three inches aperture, counted almost as many stars in that region as did Herschel with his eighteen-inch reflector. But if the stellar universe extends without limit we can hardly suppose it to do so in one plane only; hence the absence of the minuter stars and of diffused milky light over the larger part of the heavens is now held to prove that the myriads of very minute stars in the Milky Way really belong to it, and not to the depths of space far beyond.

It seems to me that here we have a fairly direct proof that the stars of our universe are really limited in number.

There are thus four distinct lines of argument all pointing with more or less force to the conclusion that the stellar universe we see around us, so far from being infinite, is strictly limited in extent and of a definite form and constitution. They may be briefly summarised as follows:—

(1) Professor Newcomb shows that, if the stars were infinite in number, and if those we see were approximately a fair sample of the whole, and further, if there were not sufficient dark bodies to shut out almost the whole of their light, then we should receive from them an amount of light theoretically greater than that of sunlight. I have shown, at some length, that neither of these causes of loss of light will account for the enormous disproportion between the

theoretical and the actual light received from the stars; and therefore Professor Newcomb's argument must be held to be a valid one against the infinite extent of our universe. Of course, this does not imply that there may not be any number of other universes in space, but as we know absolutely nothing of them—even whether they are material or non-material—all speculation as to their existence is worse than useless.

(2) The next argument depends on the fact that all over the heavens, even in the Milky Way itself, there are areas of considerable extent, besides rifts, lanes, and circular patches, where stars are either quite absent or very faint and few in number. In many of these areas the largest telescopes show no more stars than those of moderate size, while the few stars seen are projected on an intensely dark background. Sir William Herschel, Humboldt, Sir John Herschel, R.A. Proctor, and many living astronomers hold that, in these dark areas, rifts, and patches, we see completely through our stellar universe into the starless depths of space beyond.

(3) Then we have the remarkable fact that the steady increase in the number of stars, down to the ninth or tenth magnitudes, following one constant ratio either gradually or suddenly changes, so that the total number from the tenth down to the seventeenth magnitudes is only about one-tenth of what it would have been had the same ratio of increase continued. The conclusion to be drawn from this fact clearly

is, that these faint stars are becoming more and more thinly scattered in space, while the dark background on which they are usually seen shows that, except in the region of the Milky Way, there are *not* multitudes of still smaller invisible stars beyond them.

(4) The last indication of a limited stellar universe—the estimate of numbers by the light-ratio of each successive magnitude—powerfully supports the three preceding arguments.

The four distinct classes of evidence now adduced must be held to constitute, as nearly as the circumstances permit, a satisfactory proof that the stellar universe, of which our solar system forms a part, has definite limits; and that a full knowledge of its form, structure, and extent, is not beyond the possibility of attainment by the astronomers of the future.

CHAPTER VIII

OUR RELATION TO THE MILKY WAY

We now approach what may be termed the very heart of the subject of our inquiry, the determination of how we are actually situated within this vast but finite universe, and how that position is likely to affect our globe as being the theatre of the development of life up to its highest forms.

We begin with our relation to the Milky Way (which we have fully described in our fourth chapter), because it is by far the most important feature in the whole heavens. Sir John Herschel termed it 'the ground-plane of the sidereal system'; and the more it is studied the more we become convinced that the whole of the stellar universe—stars, clusters of stars, and nebulæ—are in some way connected with it, and are probably dependent on it or controlled by it. Not only does it contain a greater number of stars of the higher magnitudes than any other part of the heavens of equal extent, but it also comprises a great preponderance of star-clusters, and a great extent of diffused nebulous matter, besides the innumerable myriads of minute stars which produce its characteristic cloud-like appearance. It is also the region of those strange outbursts forming new stars;

while gaseous stars of enormous bulk—some probably a thousand or even ten thousand times that of our sun, and of intense heat and brilliancy—are more abundant there than in any other part of the heavens. It is now almost certain that these enormous stars and the myriads of minute stars just visible with the largest telescopes, are actually intermingled, and together constitute its essential features; in which case the fainter stars are really small and cannot be far apart, forming, as it were, the first aggregations of the nebulous substratum, and perhaps supplying the fuel which keeps up the intense brilliancy of the giant suns. If this is so, then the Galaxy must be the theatre of operation of vast forces, and of continuous combinations of matter, which escape our notice owing to its enormous distance from us. Among its millions of minute telescopic stars, hundreds or thousands may appear or disappear yearly without being perceived by us, till the photographic charts are completed and can be minutely scrutinised at short intervals. As undoubted changes have occurred in many of the larger nebulæ during the last fifty years, we may anticipate that analogous changes will soon be noted in the stars and the nebulous masses of the Milky Way. Dr. Isaac Roberts has even observed changes in nebulæ after such a short interval as eight years.

The Milky Way a Great Circle

Notwithstanding all its irregularities, its divisions, and its diverging branches, astronomers are generally agreed that the Milky Way forms a great circle in the heavens. Sir John Herschel, whose knowledge of it was unrivalled, stated that its course 'conforms, as nearly as the indefiniteness of its boundary will allow it to be fixed, to that of a great circle'; and he gives the Right Ascension and Declination of the points where it crosses the equinoctial, in figures which define those points as being exactly opposite each other. He also defines its northern and southern poles by other figures, so as to show that they are the poles of a great circle. And after referring to Struve's view that it was *not* a great circle, he says, 'I retain my own opinion.' Professor Newcomb says that its position 'is nearly always near a great circle of the sphere'; and again he says: 'that we are in the galactic plane itself seems to be shown in two ways: (1) the equality in the counts of stars on the two sides of this plane all the way to its poles; and (2) the fact that the central line of the Galaxy is a great circle, which it would not be if we viewed it from one side of its central plane' (*The Stars*, p. 317). Miss Clerke, in her *History of Astronomy*, speaks of 'our situation *in* the galactic plane' as one of the undisputed facts of astronomy; while Sir Norman Lockyer, in a lecture delivered in 1899, said, 'the middle line of the Milky Way is really not distinguishable from a great circle,' and again in the same lecture—'but the recent work, chiefly of Gould in Argentina, has shown that it practically is a great circle.'

About this fact, then, there can be no dispute. A great circle is a circle dividing the celestial sphere into two equal portions, as seen from the earth, and therefore the plane of this circle must pass through the earth. Of course the whole thing is on such a vast scale, the Milky Way varying from ten to thirty degrees wide, that the plane of its circular course cannot be determined with minute accuracy. But this is of little importance. When carefully laid down on a chart, as in that of Mr. Sidney Waters (see end of volume), we can see that its central line does follow a very even circular course, conforming 'as nearly as may be' to a great circle. We are therefore certainly well within the space that would be enclosed if its northern and southern margins were connected together across the vast intervening abyss, and in all probability not far from the central plane of that enclosed space.

The Form of the Milky Way and our Position on its Plane

Although the Galaxy forms a great circle in the heavens from our point of view, it by no means follows that it is circular in plan. Being unequal in width and irregular in outline, it might be elliptic or even angular in shape without being at all obviously so to us. If we were standing in an open plain or field two or three miles in diameter, and bounded in every direction by woods of very irregular height and density and great diversity of tint, we should find it difficult to judge of the shape of the field, which might be either a true circle, an oval, a hexagon, or quite irregular in outline, without our being able to detect the exact shape unless some parts were very much nearer to us than others. Again, just as the woods bounding the field might be either a narrow belt of nearly uniform width, or might in some places be only a few yards wide and in others stretch out for miles, so there have been many opinions as to the width of the Milky Way in the direction of its plane, that is, in the direction in which we look towards it. Lately, however, as the result of long-continued observation and study, astronomers are fairly well agreed as to its general form and extent, as will be seen by the following statements of fact and reasoning.

Miss Clerke, after giving the various views of many astronomers—and as the historian of modern astronomy her opinion has much weight—considers that the most probable view of it is, that it is really very much what it seems to us—an immense ring with streaming appendages extending from the main body in all directions, producing the very complex effect we see. The belief seems to be now spreading that the whole universe of stars is spherical or spheroidal, the Milky Way being its equator, and therefore in all probability circular or nearly so in plan; and it is also held that it must be rotating—perhaps very slowly—as nothing else can be supposed to have led to the formation of such a vast ring, or can preserve it when formed.

Professor Newcomb considers, from the numbers of the stars in all directions towards the Milky Way being approximately equal, that there cannot be much difference in our distance from it in various directions. It would follow that its plan is approximately circular or broadly elliptic. The existence of ring-nebulæ may be held to render such a form probable.

Sir Norman Lockyer gives facts which tend in the same direction. In an article in *Nature* of November 8th, 1900, he says: 'We find that the gaseous stars are not only confined to the Milky Way, but they are the most remote in every direction, in every galactic longitude; all of them have the smallest proper motion.' And again, referring to the hottest stars being equally remote on all sides of us, he says: 'It is because we are in the centre, because the solar system is in the centre, that the observed effect arises.' He also considers that the ring-nebula in Lyra nearly represents the form of our whole system; and he adds: 'We practically know that in our system the centre is the region of least disturbance, and therefore cooler conditions.'

These various facts and conclusions of some of the most eminent astronomers all point to one definite inference, that our position, or that of the solar system, is not very far from the centre of the vast ring of stars constituting the Milky Way, while the same facts imply a nearly circular form to this ring. Here, more than as regards our position in the plane of the Galaxy, there is no possibility of precise determination; but it is quite certain that if we were situated very far away from the centre, say, for instance, one-fourth of its diameter from one side of it and three-fourths from the other, the appearances would not be what they are, and we should easily detect the excentricity of our position. Even if we were one-third the diameter from one side and two-thirds from the other, it will, I think, be admitted that this also would have been ascertained by the various methods of research now available. We must, therefore, be somewhere between the actual centre and a circle whose radius is one-third of the distance to the Milky Way. But if we are about midway between these two positions, we shall only be one-sixth of the radius or one-twelfth of the diameter of the Milky Way from its exact centre; and if we form part of a cluster or group of stars slowly revolving around that centre, we should probably obtain all the advantages, if any, that may arise from a nearly central position in the entire star-system.

This question of our situation within the great circle of the Milky Way is of considerable importance from the point of view I am here suggesting, so that every fact bearing upon it should be noted; and there is one which has not, I think, been given the full weight due to it. It is generally admitted that the greater brilliancy of some parts of the Milky Way is no indication of nearness, because surfaces possess equal brilliancy from whatever distance they are seen. Thus each planet has its special brilliancy or reflective power, technically termed its 'albedo,' and this remains the same at all distances if the other conditions are similar. But notwithstanding this well-known fact, Sir John Herschel's remark that the greater brightness of the southern Milky Way 'conveys strongly the impression of greater proximity,' and therefore, that we are excentrically placed in its plane, has been adopted by many writers as if it were the statement of a fact, or at least a clearly expressed opinion, instead of being a mere 'impression,' and really a misleading one. I therefore wish to adduce a phenomenon which has a real bearing on the question. It is evident that, if the Milky Way were actually of uniform width throughout, then differences of apparent width would indicate differences of distance. In the parts nearer to us it would appear wider, where more remote, narrower; but in these opposite directions there would not necessarily be any differences in brightness. We should, however, expect that in the parts nearer to us the lucid stars, as well as those within any definite limits of magnitude, would be either more numerous or more wide apart on the average. No such difference as this, however, has been recorded; but there *is* a peculiar correspondence in the opposite portions of the Galaxy which is very suggestive. In the beautiful charts of the Nebulæ and Star Clusters by the late Mr. Sidney Waters, published by the Royal Astronomical Society and here reproduced by their permission (see end of volume), the Milky Way is delineated in its whole extent with great detail and from the best authorities. These charts show us that, in both hemispheres, it reaches its maximum extension on the right and left margins of the charts, where it is almost equal in extent; while in the centre of each chart, that is at its nearest points to the north and south poles respectively, it is at its narrowest portion; and, although this part in the southern hemisphere is brightest and most strongly defined, yet the actual extent, including the fainter portions, is, again, not very unequal in the opposite segments. Here we have a remarkable and significant symmetry in the proportions of the Milky Way, which, taken in connection with the nearly symmetrical scattering of the stars in all parts of the vast ring, is strongly suggestive of a nearly circular form and of our nearly central position within its plane. There is one other feature in this delineation of the Milky Way which is worthy of notice. It has been the universal practice to speak of it as being double through a considerable portion of its extent, and all the usual star-maps show the division greatly exaggerated, especially in the northern hemisphere; and this division was considered so important as to lead to the cloven-disc theory of its form, or that it consisted of two separate irregular rings, the nearer one partly hiding the more distant; while various spiral combinations were held by others to be the best way of explaining its complex appearance. But this newer map, reduced from a large one by Lord Rosse's astronomer, Dr. Boeddicker, who devoted five years to its delineation, shows us that there is no actual division in any portion of it in the northern hemisphere, but that everywhere, throughout its whole width, it consists of numerous intermingled streams and branches, varying greatly in luminosity, and with many faint or barely distinguishable extensions along its margins, yet forming one unmistakable nebulous belt; and the same general character applies to it in the southern hemisphere as delineated by Dr. Gould.

Another feature, which is well shown to the eye by these more accurate maps, is the regular curvature of the central line of the Milky Way. We can judge of this almost sufficiently by the eye; but if, with a pair of compasses, we find the proper radius and centre of curvature, we shall see that the true circular curve is always in the very centre of the nebulous mass, and the same radius applied in the same manner to the opposite hemisphere gives a similar result. It will be noted that as the Milky Way is obliquely situated on these charts, the centre of the curve will be about in R.A. 0h. 40m. in the map of the southern hemisphere, and in

R.A. 12h. 40m. in that of the northern hemisphere; while the radius of curvature will be about the length of the chord of eight hours of R.A. as measured on the margin of the maps. This great regularity of curve of the central line of the Galaxy strongly suggests rotation as the only means by which it could have originated and be maintained.

The Solar Cluster

Astronomers are now generally agreed that there is a cluster of stars of which our sun forms a part, though its exact dimensions, form, and limits are still under discussion. Sir William Herschel long ago arrived at the conclusion that the Milky Way 'consists of stars very differently scattered from those immediately around us.' Dr. Gould believed that there were about five hundred bright stars much nearer to us than the Milky Way, which he termed the solar cluster. And Miss Clerke observes that the actual existence of such a cluster is indicated by the fact that 'an enumeration of the stars in photometric order discloses a systematic excess of stars brighter than the 4th magnitude, making it certain that there is an actual condensation in the neighbourhood of the sun—that the average allowance of cubical space per star is smaller within a sphere enclosing him with a radius, say, of 140 light-years, than further away.'

But the most interesting inquiry into this subject is that by Professor Kapteyn of Gröningen, one of the most painstaking students of the distribution of the stars. He founds his conclusions mainly on the proper motions of the stars, this being the best general indication of distance in the absence of actual determination of parallax. He made use of the proper motions and the spectra of more than two thousand stars, and he finds that a considerable body of stars having large proper motions, and also presenting the solar type of spectra, surround our sun in all directions, and show no increased density, as the more distant stars do, towards the Milky Way. He finds also that towards the centre of this cluster stars are far closer together than near its outer limits (he says there are ninety-eight times as many), that it is roughly spherical in shape, and that the maximum compression is, as nearly as can be ascertained, at the centre of the circle of the Milky Way, while the sun is at some distance away from this central point.

It is a very suggestive fact that most of the stars belonging to this cluster have spectra of the solar type, which indicates that they are of the same general chemical constitution as our sun, and are also at about the same stage of evolution; and this may well have arisen from their origin in a great nebulous mass situated at or near the centre of the galactic plane, and probably revolving round their common centre of gravity.

As Kapteyn's result was based on materials which were not so full or reliable as those now available, Professor S. Newcomb has examined the question himself, using two recent lists of stars, one limited to those having proper motions of 10" a century, of which there are 295, and the other of nearly 1500 stars with 'appreciable proper motions.' They are situated in two zones, each about 5° in breadth and cutting across the Milky Way in different parts of its course. They afford, therefore, a good test of the distribution of these nearer stars with regard to the Galaxy. The result is, that on the average these stars are not more numerous in or near the Milky Way than elsewhere; and Professor Newcomb expresses himself on this point as follows:—'The conclusion is interesting and important. If we should blot out from the sky all the stars having no proper motion large enough to be detected, we should find remaining stars of all magnitudes; but they would be scattered almost uniformly over the sky, and show little or no tendency to crowd towards the Galaxy, unless, perhaps, in the region near 19h. of Right Ascension.'

A little consideration will show that, as the stars of all magnitudes which are, on the average, nearest to us are spread over the sky in 'all directions' and 'almost uniformly,' this necessarily implies that they form a cluster or group, and that our sun is somewhere not very far from the centre of this group. Again, Professor Newcomb refers to 'the remarkable equality in the number of stars in opposite directions from us. We do not detect any marked difference between the numbers lying round the opposite poles of the Galaxy, nor, so far as known, between the star-density in different regions at equal distances from the Milky Way' (*The Stars*, p. 315). And again he refers to the same question at p. 317, where he says: 'So far as we can judge from the enumeration of the stars in all directions, and from the aspect of the Milky Way, our system is near the centre of the stellar universe.'

It will, I think, now be clear to my readers that the four main astronomical propositions stated in my article which appeared in the New York *Independent* and in the *Fortnightly Review*, and which were either denied or declared to be unproved by my astronomical critics, have been shown to be supported by so many converging lines of evidence, that it is no longer possible to deny that they are, at least provisionally, fairly well established. These facts are, (1) that the stellar universe is not of infinite extent; (2) that our sun is situated in the central plane of the Milky Way; (3) that it is also situated near to the centre of that plane; (4) that we are surrounded by a group or cluster of stars of unknown extent, which occupy a place not far removed from the centre of the galactic plane, and therefore, near to the centre of our universe of stars.

Not only are these four propositions each supported by converging lines of evidence, including some which I believe have not before been adduced in their support, but a number of astronomers, admittedly of the first rank, have arrived at the same conclusions as to the bearing of the evidence, and have expressed their convictions in the clearest manner, as quoted by me. It is *their* conclusions which I appeal to and adopt; yet my two chief astronomical critics positively deny that there is any valid evidence of the finiteness of the stellar universe, which one of them terms 'a myth,' and he even accuses *me* of having started it. Both of them, however, agree in stating very strongly one objection to my main thesis— that our central position (not necessarily at the precise centre) in the stellar universe has a meaning and a purpose, in connection with the development of life and of man upon this earth, and, *so far as we know*, here only. With this one objection, the only one that in my opinion has the slightest weight, I will now proceed to deal.

The Sun's Motion through Space

The two astronomers who did me the honour to criticise my original article laid the greatest stress on the fact, that even if I had proved that the sun now occupied a nearly central position in the great star-system, it was really of no importance whatever, because, at the rate the sun was travelling, 'five million years ago we were deep in the actual stream of the Milky Way; five million years hence we shall have completely crossed the gulf which it encircles, and again be a member of one of its constituent groups, but on the opposite side. And ten million years are regarded by geologists and biologists as but a trifle on account to meet their demands upon the bank of Time.' Thus speaks one of my critics. The other is equally crushing. He says:—'If there is a centre to the visible universe, and if we occupy it to-day, we certainly did not do so yesterday, and shall not do so to-morrow. The Solar System is known to be moving among the stars with a velocity which would carry us to Sirius within 100,000 years, if we happened to be travelling in his direction, as we are not. In the 50 or 100 million years during which, according to geologists, this earth has been a habitable globe, we must have passed by thousands of stars on the right hand and on the left.... In his eagerness to limit the universe in space, Dr. Wallace has surely forgotten that it is equally important, for his purpose, to limit it in time; but incomparably more difficult in the face of ascertained facts.... Indeed, so far from our having tranquilly enjoyed a central position in unbroken continuity for scores or perhaps hundreds of millions of years, we should in that time have traversed the universe from boundary to boundary.'

Now the average reader of these two criticisms, taking account of the high official position of both writers, would accept their statements of the case as being demonstrated facts, requiring no qualification whatever, and would conclude that my whole argument had been thereby rendered worthless, and all that I founded upon it a fantastic dream. But if, on the other hand, I can show that their stated facts as to the sun's motion are by no means demonstrated, because founded upon assumptions which may be quite erroneous; and further, that if the facts should turn out to be substantially correct, they have both omitted to state well-known and admitted qualifications which render the conclusions they derive from the facts very doubtful, then the average reader will learn the valuable lesson that official advocacy, whether in medicine, law, or science is never to be accepted till the other side of the case has been heard. Let us see, therefore, what the facts really are.

Professor Simon Newcomb calculates that, if there are one hundred million stars in the stellar universe each five times the mass of our sun, and spread over a space which light would require thirty thousand years to cross, then any mass traversing such a system with a velocity of more than twenty-five miles a second, would fly off into infinite space never to return. Now as there are many stars which have, apparently, very much more than this velocity, it would follow that the visible universe is unstable. It also implies that these great velocities were not acquired in the system itself, but that the bodies which possess them must have entered it from without, thus requiring other universes as the feeders of our universe.

For the accuracy of the above statement the authority of Professor Newcomb is an ample guarantee; but there may be modifications required in the data on which it is founded, and these may greatly alter the result. If I do not mistake, the estimate of a hundred million stars is founded on actual counts or estimates of stars of successive magnitudes in different parts of the heavens, and it does not include either those of the denser star clusters nor the countless millions just beyond the reach of telescopes in the Milky Way. Neither does it make allowance for the dark stars supposed by some astronomers to be many times more numerous than the bright ones, nor for the vast number of the nebulæ, great and small, in calculating the total mass of the stellar system. In his latest work Professor Newcomb says, 'The total number of stars is to be counted by hundreds of millions'; and hence the controlling power of the system on bodies within it will be many times greater than that given above, and might even be ample to retain within its bounds such a rapidly moving star as Arcturus, which is believed to be travelling at the rate of more than three hundred miles a second. But there is another very important limitation to the conclusions to be drawn from Professor Newcomb's calculation. It assumes the stars to be nearly uniformly distributed through the whole of the space to which the system extends. But the facts are very different. The existence of clusters, some of which comprise many thousands of stars, is one example of irregularity of distribution, and anyone of these larger clusters would probably be able to change the course of even the swiftest stars passing near it. The larger nebulæ might have the same effect, since the late Mr. Ranyard, taking all his data so as to produce a minimum result, calculated the probable mass of the Orion nebula to be four and a half million times that of the sun, and there may be many other nebulæ equally large. But far more important is the fact of the vast ring of the Milky Way, which is now universally held by astronomers to be, not only apparently but really, more densely crowded with stars and also with vast masses of nebulous matter than any other part of the heavens, so that it may possibly comprise within itself a very large proportion of the whole of the matter of the visible universe. This is rendered more probable by the fact that the great majority of star-clusters lie along its course, most of the huge gaseous stars belong to it, while the occurrence there only of 'new stars' is evidence of a superabundance of matter in various forms leading to frequent heat-producing collisions, just as the frequent occurrence of meteoric showers on our earth is evidence of the superabundance of meteoric matter in the solar system.

It is recognised by mathematicians that within any great system of bodies subject to the law of gravitation there can be no such thing as motion of any of them in a straight line; neither can any amount of motion arise within such a system through the action of gravitation alone capable of carrying any of its masses out of the system. The ultimate tendency must be towards concentration rather than towards dispersal.

It seems, therefore, only reasonable to consider whatever motions and whatever velocities we find among the stars, as having been produced by the gravitative power of the larger aggregations, modified perhaps by electrical repulsive forces, by collisions, and by the results of those collisions; and we may look to the changes now visibly going on in some of the nebulæ and clusters as indications of the forces that have probably brought about the actual condition of the whole stellar universe.

If we examine the beautiful photographs of nebulæ by Dr. Roberts and other observers, we find that they are of many forms. Some are extremely irregular and almost like patches of cirrus clouds, but a large number are either distinctly spiral in form, or show indications of becoming spiral, and this has been found to be the case even with some of the large irregular nebula. Then again we have numerous ring-formed nebulæ, usually with a star involved in dense nebulosity in the centre, separated by a dark space of various widths from the outer ring. All these kinds of nebulæ have stars involved in them, and apparently forming part of their structure, while others which do not differ in appearance from ordinary stars are believed by Dr. Roberts to lie between us and the nebula. In the case of many of the spiral nebulæ, stars are often strung along the coils of the spiral, while other curved lines of stars are seen just outside the nebula, so that it is impossible to avoid the conclusion that both are really connected with it, the outer lines of stars indicating a former greater extension of the nebula whose material has been used up in the growth of these stars. Some of these spiral nebulæ show beautifully regular convolutions, and these usually have a large central star like mass, as in M. 100 Comæ and I. 84 Comæ, in Vol. II. Pl. 14 of Dr. Roberts's photographs. The straight white streaks across the nebula of the Pleiades and some others are believed by Dr. Roberts to be indications of spiral nebulæ seen edgewise. In other cases, clusters of stars are more or less nebulous, and the arrangement of the stars seems to indicate their development from a spiral nebula. It is to be noted that many of the objects classed as planetary nebulæ by Sir John Herschel are shown by the best photographs to be really of the ring-type, though often with a very narrow division between the ring and the central mass. This form may therefore be of frequent occurrence.

But if this annular form with some kind of central nucleus, often very large, is produced under certain conditions by the action of the ordinary laws of motion upon more or less extensive masses of discrete matter, why may not the same laws acting upon similar matter once dispersed over the whole extent of the existing stellar universe, or even beyond what are now its farthest limits, have led to the aggregation of the vast annular formation of the Milky Way, with all the subordinate centres of concentration or dispersal to be found within or around it? And if this is a reasonable conception, may we not hope that by a concentration of attention upon a few of the best marked and most favourably situated annular and spiral systems, sufficient knowledge of their internal motions may be obtained which may serve as a guide to the kind of motion we may expect to find in the great galactic ring and its subordinate stars? We may then perhaps discover which now seem so erratic, are really all parts of a series of orbital movements limited and controlled by the forces of the great system to which they belong, so that, if not mathematically stable, they may yet be sufficiently so to endure for some thousand millions of years.

It is a suggestive fact that the calculated position of the 'solar apex'—the point towards which our sun appears to move—is now found to be much more nearly in the plane of the Milky Way than the position first assigned to it, and Professor Newcomb adopts, as most likely to be accurate, a point near the bright star Vega in the constellation Lyra. Other calculators have placed it still farther east, while Rancken and Otto Stumpe assign it a position actually in the Milky Way; and Mr. G.C. Bompas concludes that the sun's plane of motion nearly coincides with that of the

Galaxy. M. Rancken found that 106 stars near the Milky Way showed, in their very small proper motions, a drift along it in a direction from Cassiopeiæ towards Orion, and this, it is supposed, may be partly due to our sun's motion in an opposite direction.

In many other parts of the heavens there are groups of stars which have almost identical proper motions—a phenomenon which the late R.A. Proctor termed 'star-drift'; and he especially pointed out that five of the stars of the Great Bear were all drifting in the same direction; and although this has been denied by later writers, Professor Newcomb, in his recent book on *The Stars*, declares that Proctor was right, and explains that the error of his critics was due to not making allowance for the divergence of the circles of right ascension. The Pleiades are another group, the stars of which drift in the same direction, and it is a most suggestive fact that photographs now show this cluster to be embedded in a vast nebula, which, therefore, has also a proper motion; but some of the smaller stars do not partake of it. Three stars in Cassiopeiæ also move together, and no doubt many other similarly connected groups remain to be discovered.

These facts have a very important bearing on the question of the motion of our sun in space. For this motion has been determined by comparing the motions of large numbers of stars which are assumed to be wholly independent of each other, and to move, as it were, at random. Miss A.M.

Clerke, in her *System of the Stars*, puts this point very clearly, as follows: 'For the assumption that the absolute movements of the stars have no preference for one direction over another, forms the basis of all investigations hitherto conducted into the translatory advance of the solar system. The little fabric of laboriously acquired knowledge regarding it at once crumbles if that basis has to be removed. In all investigations of the sun's movement, the movements of the stars have been regarded as casual irregularities; should they prove to be in any visible degree systematic, the mode of treatment adopted (and there is no other at present open to us) becomes invalid, and its results null and void. The point is then of singular interest, and the evidence bearing upon it deserves our utmost attention.'

Mr. W.H.S. Monck, a well-known astronomer, takes the same view. He says: 'The proof of this motion rests on the assumption that if we take a sufficient number of stars, their real motions in all directions will be equal, and that therefore the apparent preponderances which we observe in particular directions result from the real motion of the sun. But there is no impossibility in a systematic motion of the majority of the stars used in these researches which might reconcile the observed facts with a motionless sun. And, in the second place, if the sun is not in the exact centre of gravity of the universe, we might expect him to be moving in an orbit around this centre of gravity, and our observations on his actual motion are not sufficiently numerous or accurate to enable us to affirm that he is moving in a right line rather than such an orbit.'

Now this 'systematic motion,' which would render all calculations as to the sun's motion inaccurate or even altogether worthless, is by many astronomers held to be an observed reality. The star-drift, first pointed out by Proctor, has been shown to exist in many other groups of stars, while the curious arrangements of stars all over the heavens in straight lines, or regular curves, or spirals, strongly suggests a wide extension of the same kind of relation. But even more extensive systematic movements have been observed or suggested by astronomers. Sir D. Gill, by an extensive research, believes that he has found indications of a rotation of the brighter fixed stars as a whole in regard to the fainter fixed stars as a whole. Mr. Maxwell Hall has also found indications of a movement of a large group of stars, including our sun, around a common centre, situated in a direction towards Epsilon Andromedæ, and at a distance of about 490 years of light-travel. These last two motions are not yet established; but they seem to prove two important facts— (*a*) that eminent astronomers believe that *some* systematic motions must exist among the stars, or they would not devote so much labour to the search for them; and (*b*) that extensive systematic motions of some kind do exist, or even these results would not have been obtained.

Mr. W.W. Campbell, of the Lick Observatory, thus remarks on the uncertainty of determinations of the sun's motions: 'The motion of the solar system is a purely relative quantity. It refers to specified groups of stars. The results for various groups may differ widely, and all be correct. It would be easy to select a group of stars with reference to which the solar motion would be reversed 180° from the values assigned above' (*Astrophysical Journal*, vol. xiii. p. 87. 1901).

It must be remembered that, within a uniform cluster of stars, each moving round the common centre of gravity of the whole cluster, Kepler's laws do not prevail, the law being that the angular velocities are all identical, so that the more distant stars move faster than those nearer the centre, subject to modifications, however, due to the varying density of the cluster. But if the cluster is nearly globular, there must be stars moving round the centre in every plane, and this would lead to apparent motions in many directions as viewed by us, although those which were moving in the same plane as ourselves would, when compared with remote stars outside the cluster, appear to be all moving in the same direction and at the same rate, forming, in fact, one of those drifting systems of stars already referred to. Again, if in the process of formation of our cluster, smaller aggregations already having a rotatory motion were drawn into it, this might lead to their revolving in an opposite direction to those which were formed from the original nebula, thus increasing the diversities of apparent motion.

The evidence now briefly set forth fully justifies, I submit, the remarks as to the statements of my astronomical critics at the beginning of this section. They have both given the accepted views as to direction and rate of movement of our sun without any qualification whatever, as if they were astronomical facts of the same certainty and the same degree of accuracy as the sun's distance from the earth; and they will assuredly have been so understood by the great body of non-mathematical readers. It appears, however, if the authorities I have quoted are right, that the whole calculation rests upon certain assumptions, which are certainly to some extent, and may be to a very large extent, erroneous. This is my reply to one part of their criticism.

In the next place, they both assert, or imply, not only that the sun's motion is now in a straight line, but that it has been in a straight line from some enormously remote period when it first entered the stellar system on one side, and will so continue to move till it reaches the utmost bounds of that system on the other side. And this is stated by them both, not as a possibility, but as a certainty. They use such terms as 'must' and 'will be,' leaving no room for any doubt whatever. But such a result implies the abrogation of the law of gravitation, since under its action motion in a straight line in the midst of thousands or millions of suns of various sizes is an absolute impossibility; while it also implies that the sun must have been started on its course from some other system outside the Milky Way, with such a precise determination of direction as not to collide with, or even make a near approach to, any one of the suns or clusters of suns, or vast nebulous masses, during its passage through the very midst of the stellar universe.

This is my reply to the main point of their criticism, and I think I am justified in saying that nothing in my whole article is so demonstrably baseless as the statements I have now examined.

Considering then the whole bearing of the evidence, I refuse to accept the unsupported dicta of those who would have us believe that our admitted position not far from the centre of the stellar universe is a mere temporary coincidence of no significance whatever; or that our sun and hosts of other similar orbs near to us have come together by an accident, and are being dispersed into surrounding space, never to meet again. Until this is proved by indisputable evidence, it seems to me far more probable that we are moving in an orbit of some kind around the centre of gravity of a vast cluster, as determined by the investigations of Kapteyn, Newcomb, and other astronomers; and, consequently, that the nearly central position we now occupy may be a permanent one. For even if our sun's orbit should have a diameter a thousand times that of Neptune, it would be but a small fraction of the diameter of the Milky Way; while so vast is the scale of our universe, that it might be even a hundred thousand times as great and still leave us deeply immersed in the solar cluster, and very much nearer to the dense central portion than to its more diffused outer regions.

Here the subject may be left for the present. After having studied the evidence afforded by the essential conditions of life-development on the earth, and the numerous indications that these conditions do not exist on any of the other planets of the solar system, it may be again touched upon in a general review of the conclusions arrived at.

CHAPTER IX

THE UNIFORMITY OF MATTER AND ITS LAWS THROUGHOUT

THE STELLAR UNIVERSE

I have shown in the second chapter of this work that none of the previous writers on the question of the habitability of the other planets have really dealt with the subject in any adequate manner, since not only do they appear to be quite unaware of the delicate balance of conditions which alone renders organic life possible on any planet, but they have altogether omitted any reference to the fact that not only must the conditions be such as to render life possible *now*, but these conditions must have persisted during the long geological epochs needed for the slow development of life from its most rudimentary forms. It will therefore be necessary to enter into some details both as to the physical and chemical essentials for a continuous development of organic life, and also into the combination of mechanical and physical conditions which are required on any planet to render such life possible.

The Uniformity of Matter

One of the most important and far-reaching of the discoveries due to the spectroscope is that of the wonderful identity of the elements and material compounds in earth and sun, stars and nebulæ, and also of the identity of the physical and chemical laws that determine the states and forms assumed by matter. More than half the total number of the known elements have been already detected in the sun, including all those which compose the bulk of the earth's solid material, with the one exception of oxygen. This is a very large proportion when we consider the very peculiar conditions which enable us to detect them. For we can only recognise an element in the sun when it exists at its surface in an incandescent state, and also above its surface in the form of a somewhat cooler gas. Many of the elements may rarely or never be brought to the surface of so vast a body, or if they do sometimes appear there, it may not be in sufficient quantity or in sufficient purity to produce any bands in the spectroscope, while the cooler gas or vapour may either not be present, or be so dispersed as not to produce sufficient absorption to render its spectral lines visible. Again, it is believed that many elements are dissociated by the intense heat of the sun, and may not be recognisable by us, or they may only exist at its surface in a compound form unknown on the earth; and in some such way those lines of the solar spectrum which remain still unrecognised may have been produced. One of these unknown lines was that of Helium, a gas found soon afterwards in the rare mineral 'Cleveite,' and since detected frequently in many stars. Some of the stars have spectra very closely resembling that of the sun. The dark lines are almost as numerous, and most of them correspond accurately with solar lines, so that we cannot doubt their having almost exactly the same chemical constitution, and being also in the same condition as regards heat and stage of development. Other stars, as we have already stated, exhibit mainly lines of hydrogen, sometimes combined with fine metallic lines. Of the spectra of the nebulæ comparatively little is known, but many are decidedly gaseous, while others show a continuous spectrum indicating a more complex constitution.

But we also obtain considerable knowledge of the matter of non-terrestrial bodies by the analysis of the numerous meteorites which fall upon the earth. Most of these belong to some of the many meteoric streams which circulate round the sun, and which may be supposed to give us samples of planetary matter. But as it is now believed that many of them are produced by the debris of comets, and the orbits of some of these indicate that they have come from stellar space and

have been drawn into our system by the attractive power
of the larger planets, it is almost certain that the meteoric
stones not infrequently bring us matter from the remoter
regions of space, and probably afford us samples of the solid
constituents of nebula; or the cooler stars. It is, therefore, a
most suggestive fact that none of these meteorites have been
found to contain a single non-terrestrial element, although
no less than twenty-four elements have been found in them,
and it will be of interest to give the list of these, as follows:—
Oxygen, Hydrogen, *Chlorine*, *Sulphur*, *Phosphorus*, Carbon,
Silicon, Iron, Nickel, Cobalt, Magnesium, Chromium,
Manganese, Copper, Tin, *Antimony*, Aluminium, Calcium,
Potassium, Sodium, *Lithium*, Titanium, *Arsenic*, and
Vanadium. Seven of the above, printed in italics, have
not yet been found in the sun, such as Oxygen, Chlorine,
Sulphur, and Phosphorus, which form the constituents of
many widespread minerals, and they supply important gaps
in the series of solar and stellar elements. It may be noted
that although meteorites have supplied no new elements,
they have furnished examples of some new combinations of
these elements forming minerals distinct from any found in
our rocks.

The fact of the occurrence in meteorites not only of
minerals which are peculiar to them or are found on the
earth, but also of structures resembling our breccias,
veins, and even slicken-side surfaces, has been held to be

opposed to the meteoritic theory of the origin of suns and planets, because meteorites seem to be thus proved to be the fragments of suns or worlds, not their primary constituents. But these cases are exceptional, and Mr. Sorby, who made a special study of meteorites, concluded that their materials have usually been in a state of fusion or even of vapour, as they now exist in the sun, and that they became condensed into minute globular particles, which afterwards collected into larger masses, and may have been broken up by mutual impact, and again and again become aggregated together— thus presenting features which are completely in accordance with the meteoritic theory.

But, quite recently, Mr. T.C. Chamberlin has applied the theory of tidal distortion to showing how solid bodies in space, without ever coming into actual contact, must sometimes be torn apart or disrupted into numerous fragments by passing near to each other. Especially when a small body passes near a much larger one, there is a certain distance of approach (termed the Roche limit) when the increasing differential force of gravity will be sufficient to tear asunder the smaller body and cause the fragments either to circulate around it or to be dispersed in space. In this way, therefore, those larger meteorites which exhibit planetary structure may have been produced. Of course they would rarely have been true planets attached to a sun, but more frequently some of the smaller dark suns, which may possess many of the physical characteristics of planets, and of which there may be myriads in the stellar spaces.

On the whole, then, we have positive knowledge of the existence, in the sun, stars, and planetary and stellar spaces, of such a large proportion of the elements of our globe, and so few indications of any not forming part of it, that we are justified in the statement, that the whole stellar universe is, broadly speaking, constructed of the same series of elementary substances as those we can study upon our earth, and of which the whole realm of nature, animal, vegetable, and mineral, is composed. The evidence of this identity of substance is really far more complete than we could expect, considering the very limited means of inquiry that we possess; and we shall, therefore, not be justified in assuming that any important difference exists.

When we pass from the elements of matter to the laws which govern it, we also find the clearest proofs of identity. That the fundamental law of gravitation extends to the whole physical universe is rendered almost certain by the fact that double stars move round their common centre of gravity in elliptical orbits which correspond well with both observation and calculation. That the laws of light are the same both here and in inter-planetary space is indicated by the fact that the actual measurement of the velocity of light on the earth's surface gives a result so completely identical with that prevailing to the limits of the solar system, that the measurement of the sun's distance, by means of the eclipses of Jupiter's satellites combined with the measured velocity of light, agrees almost exactly with that obtained by means of the transits of Venus, or through our nearest approach to the planets Mars or Eros.

Again, the more recondite laws of light are found to be identical in sun and stars with those observed within the narrow bounds of laboratory experiments. The minute change of position of spectral lines caused by the source of light moving towards or away from us enables us to determine this kind of motion in the most distant stars, in the planets, or in the moon, and these results can be tested by the motion of the earth either in its orbit or in its rotation; and these latter tests agree with the theoretical determination of what must occur, dependent on the wave-lengths of the different dark lines of the solar spectrum determined by measurements in the laboratory.

In like manner, minute changes in the widening or narrowing of spectral lines, their splitting up, their increase or decrease in number, and their arrangement so as to form flutings, can all be interpreted by experiments in the laboratory, showing that such phenomena are due to alterations of temperature, of pressure, or of the magnetic field, thus proving that the very same physical and chemical laws act in the same way here and in the remotest depths of space.

These various discoveries give us the certain conviction that the whole material universe is essentially one, both as regards the action of physical and chemical laws, and also in its mechanical relations of form and structure. It consists throughout of the very same elements with which we are so familiar on our earth; the same ether whose vibrations bring us light and heat, electricity and magnetism, and a whole host of other mysterious and as yet imperfectly known forces; gravitation acts throughout its vast extent; and in whatever direction and by whatever means we obtain a knowledge of the stellar universe, we find the same mechanical, physical, and chemical laws prevailing as upon our earth, so that we have in some cases been actually enabled to reproduce in our laboratories phenomena with which we had first become acquainted in the sun or among the stars.

We may therefore feel it to be an almost certain conclusion that—the elements being the same, the laws which act upon, and combine, and modify those elements being the same—organised living beings wherever they may exist in this universe must be, fundamentally, and in essential nature, the same also. The outward forms of life, if they exist elsewhere, may vary almost infinitely, as they do vary on the earth; but, throughout all this variety of form—from fungus or moss to rose-bush, palm or oak; from mollusc, worm, or butterfly to humming-bird, elephant, or man— the biologist recognises a fundamental unity of substance and of structure, dependent on the absolute requirements of the growing, moving, developing, living organism, built up of the same elements, combined in the same proportions, and subject to the same laws. We do not say that organic life *could* not exist under altogether diverse conditions from those which we know or can conceive, conditions which may prevail in other universes constructed quite differently from ours, where other substances replace the matter and ether of our universe, and where other laws prevail. But, *within* the universe we know, there is not the slightest reason to suppose organic life to be possible, except under the same general conditions and laws which prevail here. We will, therefore, now proceed to describe, very generally, what are the conditions essential to the existence and the continuous development of vegetable and animal life.

CHAPTER X

THE ESSENTIAL CHARACTERS OF THE LIVING ORGANISM

Before trying to comprehend the physical conditions on any planet which are essential for the development and maintenance of a varied and complex system of organic life comparable to that of our earth, we must obtain some knowledge of what life is, and of the fundamental nature and properties of the living organism.

Physiologists and philosophers have made many attempts to define 'life,' but in most cases in aiming at absolute generality they have been vague and uninstructive. Thus De Blainville defined it as 'The twofold internal movement of composition and decomposition, at once general and continuous'; while Herbert Spencer's latest definition was 'Life is the continuous adjustment of internal relations to external relations.' But neither of these is sufficiently precise, explanatory, or distinctive, and they might almost be applied to the changes occurring in a sun or planet, or to the elevation and gradual formation of a continent. One of the oldest definitions, that of Aristotle, seems to come nearer the mark: 'Life is the assemblage of the operations of nutrition, growth, and destruction.' But these definitions of 'life' are unsatisfactory, because they apply to an abstract idea rather than to the actual living organism. The marvel and mystery of life, as we know it, resides in the body which manifests it, and this living body the definitions ignore.

The essential points in the living body, as seen in its higher developments, are, first, that it consists throughout of highly complex but very unstable forms of matter, every particle of which is in a continual state of growth or decay; that it absorbs or appropriates dead matter from without; takes this matter into the interior of its body; acts upon it mechanically and chemically, rejecting what is useless or hurtful; and so transforming the remainder as to renew every atom of its own structure internal and external, at the same time throwing off, particle by particle, all the worn-out or dead portions of its own substance. Secondly, in order to be able to do all this, its whole body is permeated throughout by branching vessels or porous tissues, by which liquids and gases can reach every part and carry on the various processes of nutrition and excretion above referred to. As Professor Burdon Sanderson well puts it: 'The most distinctive peculiarity of living matter as compared with non-living is, that it is ever changing while ever the same.' And these changes are the more remarkable because they are accompanied, and even produced, by a very large amount of mechanical work—in animals by means of their normal activities in search of food, in assimilating that food, in continually renewing and building up their whole organism, and in many other ways; in plants by building up their structure, which often involves raising tons of material high into the air, as in forest trees. As a recent writer puts it: 'The most prominent, and perhaps the most fundamental, phenomenon of life is what may be described as the *Energy Traffic* or the function of *trading in energy*. The chief physical function of living matter seems to consist in absorbing energy, storing it in a higher potential state, and afterwards partially expending it in the kinetic or active form.'

Thirdly—and perhaps most marvellous of all—all living organisms have the power of reproduction or increase, in the lowest forms by a process of self-division or 'fission,' as it is termed, in the higher by means of reproductive cells, which, though in their earliest stage quite indistinguishable physically or chemically in very different species, yet possess the mysterious power of developing a perfect organism, identical with its parents in all its parts, shapes, and organs, and so wonderfully resembling them, that the minutest distinctive details of size, form, and colour, in hair or feathers, in teeth or claws, in scales, spines, or crests, are reproduced with very close accuracy, though often involving metamorphic changes during growth of so strange a nature that, if they were not familiar to us but were narrated as occurring only in some distant and almost inaccessible region, would be treated as travellers' tales, incredible and impossible as those of Sindbad the Sailor.

In order that the substance of living bodies should be able to undergo these constant changes while preserving the same form and structure in minute details—that they should be, as it were, in a constant state of flux while remaining sensibly unchanged, it is necessary that the molecules of which they are built up should be so combined as to be easily separated and as easily united—be, as it is termed, *labile* or flowing; and this is brought about by their chemical composition, which, while consisting of few elements, is yet highly complex in structure, a large number of chemical atoms being combined in an endless variety of ways.

The physical basis of life, as Huxley termed it, is protoplasm, a substance which consists essentially of only four common elements, the three gases, nitrogen, hydrogen, and oxygen, with the non-metallic solid, carbon; hence all the special products of plants and animals are termed carbon-compounds, and their study constitutes one of the most extensive and intricate branches of modern chemistry. Their complexity is indicated by the fact that the molecule of sugar contains 45, and that of stearine no less than 173, constituent atoms. The chemical compounds of carbon are far more numerous than those of all the other chemical elements combined; and it is this wonderful variety and the complexity of its possible combinations which explain the fact, that all the various animal tissues—skin, horn, hair, nails, teeth, muscle, nerve, etc., consist of the same four elements (with occasionally minute quantities of sulphur, phosphorus, lime, or silica, in some of them), as proved by the marvellous fact that these tissues are all produced as well by the grass-eating sheep or ox as by the fish or flesh-eating seal or tiger. And the marvel is still further increased when we consider that the innumerable diverse substances produced by plants and animals are all formed out of the same three or four elements. Such are the endless variety of organic acids, from prussic acid to those of the various fruits; the many kinds of sugars, gums, and starches; the number of different kinds of oil, wax, etc.; the variety of essential oils which are mostly forms of turpentines, with such substances as camphor, resins, caoutchouc, and gutta-percha; and the extensive series of vegetable alkaloids, such as nicotine from tobacco, morphine from opium, strychnine, curarine, and other poisons; quinine, belladonna, and similar medicinal alkaloids; together with the essential principles of our refreshing drinks, tea, coffee, and cocoa, and others too numerous to be named here—all alike consisting solely of the four common elements from which almost our whole organism is built up. If this were not indisputably proved, it would scarcely be credited.

Professor F.J. Allen considers that the most important

element in protoplasm, and that which confers upon it its most essential properties in the living organism—its extreme mobility and transposibility—is nitrogen. This element, though inert in itself, readily enters into compounds when energy is supplied to it, the most striking illustration of which is the formation of ammonia, a compound of nitrogen and hydrogen, produced by electric discharges through the atmosphere. Ammonia, and certain oxides of nitrogen produced in the atmosphere in the same way, are the chief sources of the nitrogen assimilated by plants, and through them by animals; for although plants are continually in contact with the free nitrogen of the atmosphere, they are unable to absorb it. By their leaves they absorb oxygen and carbon-dioxide to build up their woody tissues, while by their roots they absorb water in which ammonia and oxides of nitrogen are dissolved, and from these they produce the protoplasm which builds up the whole substance of the animal world. The energy required to produce these nitrogen-compounds is given up by them when undergoing further changes, and thus the production of ammonia by electricity in the atmosphere, and its being carried by rain into the soil, constitute the first steps in that long series of operations which culminates in the production of the higher forms of life.

But the remarkable transformations and combinations continually going on in every living body, which are, in fact, the essential conditions of its life, are themselves dependent on certain physical conditions which must be always present. Professor Allen remarks: 'The sensitiveness of nitrogen, its proneness to change its state of combination and energy, appear to depend on certain conditions of temperature, pressure, etc., which exist at the surface of this earth. Most vital phenomena occur between the temperature of freezing water and 104° F. If the general temperature of the earth's surface rose or fell 72° F. (a small amount relatively), the whole course of life would be changed, even perchance to extinction.'

Another important, and even more essential fact, in connection with life, is the existence in the atmosphere of a small but nearly constant proportion of carbonic acid gas, this being the source from which the whole of the carbon in the vegetable and animal kingdoms is primarily derived. The leaves of plants absorb carbonic acid gas from the atmosphere, and the peculiar substance, chlorophyll, from which they derive their green colour, has the power, under the influence of sunlight, to decompose it, using the carbon to build up its own structure and giving out the oxygen. In the laboratory the carbon can only be separated from the oxygen by the application of heat, under which certain metals burn by combining with the oxygen, thus setting free the carbon. Chlorophyll has a highly complex chemical structure very imperfectly known, but it is said to be only produced when there is iron in the soil.

The leaves of plants, so often looked upon as mere ornamental appendages, are among the most marvellous structures in living organisms, since in decomposing carbonic acid at ordinary temperatures they do what no other agency in nature can perform. In doing this they utilise a special group of ether-waves which alone appear to have this power. The complexity of the processes going on in leaves is well indicated in the following quotation:—

'We have seen how green leaves are supplied with gases, water, and dissolved salts, and how they can trap special ether-waves. The active energy of these waves is used to transmute the simple inorganic compounds into complex organic ones, which in the process of respiration are reduced to simpler substances again, and the potential energy transformed into kinetic. These metabolic changes take place in living cells full of intense activities. Currents course through the protoplasm and cell-sap in every direction, and between the cells which are also united by strands of protoplasm. The gases used and given off in respiration and assimilation are floated in and out, and each protoplasm particle burned or unburned is the centre of an area of disturbance. Pure protoplasm is influenced equally by all rays: that associated with chlorophyll is affected by certain red and violet rays in particular. These, especially the red ones, bring about the dissociation of the elements of the carbonic acid, the assimilation of the carbon, and the excretion of the oxygen.'

It is this vigorous life-activity ever at work in the leaves, the roots, and the sap-cells, that builds up the plant, in all its wondrous beauty of bud and foliage, flower and fruit; and at the same time produces, either as useful or waste-products, all that wealth of odours and flavours, of colours and textures, of fibres and varied woods, of roots and tubers, of gums and oils and resins innumerable, that, taken altogether, render the world of vegetable life perhaps more varied, more beautiful, more enjoyable, more indispensable to our higher nature than even that of animals. But there is really no comparison between them. We *could* have plants without animals; we could *not* have animals without plants. And all this marvel and mystery of vegetable life, a mystery which we rarely ponder over because its effects are so familiar, is usually held to be sufficiently explained by the statement that it is all due to the special properties of protoplasm. Well might Huxley say, that protoplasm is not only a substance but a structure or mechanism, a mechanism kept at work by solar heat and light, and capable of producing a thousand times more varied and marvellous results than all the human mechanism ever invented.

But besides absorbing carbonic acid from the atmosphere, separating and utilising the carbon and giving out the oxygen, plants as well as animals continually absorb oxygen from the atmosphere, and this is so universally the case that oxygen is said to be the food of protoplasm, without which it cannot continue to live; and it is the peculiar but quite invisible structure of the protoplasm which enables it to do this, and also in plants to absorb an enormous amount of water as well.

But although protoplasm is so complex chemically as to defy exact analysis, being an elaborate structure of atoms built up into a molecule in which each atom must occupy its true place (like every carved stone in a Gothic cathedral), yet it is, as it were, only the starting-point or material out of which the infinitely varied structures of living bodies are formed. The extreme mobility and changeability of the structure of these molecules enables the protoplasm to be continually modified both in constitution and form, and, by the substitution or addition of other elements, to serve special purposes. Thus when sulphur in small quantities is absorbed and built into the molecular structure, proteids are formed. These are most abundant in animal structures, and give the nourishing properties to meat, cheese, eggs, and other animal foods; but they are also found in the vegetable kingdom, especially in nuts and seeds such as grain, peas, etc. These are generally known as nitrogenous foods, and are very nutritious, but not so easily digestible as meat. Proteids exist in very varied forms and often contain phosphorus as well as sulphur, but their main characteristic is the large proportion of nitrogen they contain, while many other animal and vegetable products, as most roots, tubers, and grains, and even fats and oils, are mainly composed of starch and sugar. In its chemical and physiological aspects protein is thus described by Professor W.D. Haliburton:—'Proteids are produced only in the living laboratory of animals and plants; proteid matter is the all-important material present in protoplasm. This molecule is the most complex that is known; it always contains five and often six or even seven elements. The task of thoroughly understanding its composition is necessarily vast, and advance slow. But, little by little, the puzzle is being solved, and this final conquest of organic chemistry, when it does arrive, will furnish physiologists with new light on many of the dark places of physiological science.'

What makes protoplasm and its modifications still more marvellous is the power it possesses of absorbing and moulding a number of other elements in various parts of living organisms for special uses. Such are silica in the stems of the grass family, lime and magnesia in the bones of animals, iron in blood, and many others. Besides the four elements constituting protoplasm, most animals and plants contain also in some parts of their structure sulphur, phosphorus, chlorine, silicon, sodium, potassium, calcium, magnesium, and iron; while, less frequently, fluorine, iodine, bromine, lithium, copper, manganese, and aluminium are also found in special organs or structures; and the molecules of all these are carried by the protoplasmic fluids to the places where they are required and built into the living structure, with the same precision and for similar ends as brick and stone, iron, slate, wood, and glass are each utilised in their proper places in any large building. The organism, however, is not built, but grows. Every organ, every fibre, cell, or tissue is formed from diverse materials, which are first decomposed into their elementary molecules, transformed by the protoplasm or by special solvents formed from it, carried to the places where they are needed by the vital fluids, and there built up atom by atom or molecule by molecule into the special structures of which they are to form a part.

But even this marvel of growth and repair of every individual organism is far surpassed by the greater marvel of reproduction. Every living thing of the higher orders arises from a single microscopic cell, when fertilised, as it is termed, by the absorption of another microscopic cell derived from a different individual. These cells are often, even under the highest powers of the microscope, hardly distinguishable from other cells which occur in all animals and plants and of which their structure is built up; yet these special cells begin to grow in a totally different manner, and instead of forming one particular part of the organism, develop inevitably into a complete living thing with all the organs, powers, and peculiarities of its parents, so as to be recognisably of the same species. If the simple growth of the fully formed organism is a mystery, what of this growth of thousands of complex organisms each with all its special peculiarities, yet all arising from minute germs or cells the diverse natures of which are wholly indistinguishable by the highest powers of the microscope? This, too, is said to be the work of protoplasm under the influence of heat and moisture, and modern physiologists hope some day to learn 'how it is done.' It may be well here to give the views of a modern writer on this point. Referring to a difficulty which had been stated by Clerk-Maxwell twenty-five years ago, that there was not room in the reproductive cell for the millions of molecules needed to serve as the units of growth for all the different structures in the body of the higher animals, Professor M'Kendrick says:—'But to-day, it is reasonable from existing data to suppose that the germinal vesicle might contain a million of millions of organic molecules. Complex arrangements of these molecules suited for the development of all the parts of a highly complicated organism, might satisfy all the demands of the theory of heredity. Doubtless the germ was a material system through and through. The conception of the physicist was, that molecules were in various states of movement; and the thinkers were striving toward a kinetic theory of molecules and of atoms of solid matter, which might be as fruitful as the kinetic theory of gases. There were motions atomic and molecular. It was conceivable that the peculiarities of vital action might be determined by the kind of motion that took place in the molecules of what we call living matter. It might be different in kind from some of the motions dealt with by physicists. Life is continually being created from non-living material—such, at least, is the existing view of growth by the assimilation of food. The creation of living matter out of non-living may be the transmission to the dead matter of molecular motions which are *sui generis* in form.' This is the modern physiological view of 'how it may be done,' and it seems hardly more intelligible than the very old theory of the origin of stone axes, given by Adrianus Tollius in 1649, and quoted by Mr. E.B. Tylor, who says:—'He gives drawings of some ordinary stone axes and hammers and tells how naturalists say that they are generated in the sky by a fulgureous exhalation conglobed in a cloud by the circumfixed humour, and are, as it were, baked hard by intense heat, and the weapon becomes pointed by the damp mixed with it flying from the dry part, and leaving the other end denser, but the exhalations press it so hard that it breaks through the cloud and makes thunder and lightning. But—he says—if this is really the way in which they are generated, it is odd they are not round, and that they have holes through them. It is hardly to be believed, he thinks.' And so, when the physiologists, determined to avoid the assumption of anything beyond matter and motion in the germ, impute the whole development and growth of the elephant or of man from minute cells internally alike, by means of 'kinds of motion' and the 'transmission of motions which are *sui generis* in form,' many of us will be inclined to say with the old author—'It is hardly to be believed, I think.'

This brief statement of the conclusions arrived at by chemists and physiologists as to the composition and structure of organised living things has been thought advisable, because the non-scientific reader has often no conception of the incomparable marvel and mystery of the life-processes he has always seen going on, silently and almost unnoticed, in the world around him. And this is still more the case now that two-thirds of our population are crowded into cities where, removed from all the occupations, the charms, and the interests of country life, they are driven to seek occupation and excitement in the theatre, the music-hall, or the tavern. How little do these know what they lose by being thus shut out from all quiet intercourse with nature; its soothing sights and sounds; its exquisite beauties of form and colour; its endless mysteries of birth, and life, and death. Most people give scientific men credit for much greater knowledge than they possess in these matters; and many educated readers will, I feel sure, be surprised to find that even such apparently simple phenomena as the rise of the sap in trees are not yet completely explained. As to the deeper problems of life, and growth, and reproduction, though our physiologists have learned an infinite amount of curious or instructive facts, they can give us no intelligible explanation of them.

The endless complexities and confusing amount of detail in all treatises on the physiology of animals and plants are such, that the average reader is overwhelmed with the mass of knowledge presented to him, and concludes that after such elaborate researches everything must be known, and that the almost universal protests against the need of any causes but the mechanical, physical, and chemical laws and forces are well founded. I have, therefore, thought it advisable to present a kind of bird's-eye view of the subject, and to show, in the words of the greatest living authorities on these matters, both how complex are the phenomena and how far our teachers are from being able to give us any adequate explanation of them.

I venture to hope that the very brief sketch of the subject I have been able to give will enable my readers to form some faint general conception of the infinite complexity of life and the various problems connected with it; and that they will thus be the better enabled to appreciate the extreme delicacy of those adjustments, those forces, and those complex conditions of the environment, that alone render life, and above all the grand age-long panorama of the development of life, in any way possible. It is to these conditions, as they prevail in the world around us, that we will now direct our attention.

CHAPTER XI

THE PHYSICAL CONDITIONS ESSENTIAL FOR ORGANIC LIFE

The physical conditions on the surface of our earth which appear to be necessary for the development and maintenance of living organisms may be dealt with under the following headings:—

1. Regularity of heat-supply, resulting in a limited range of temperature.

2. A sufficient amount of solar light and heat.

3. Water in great abundance, and universally distributed.

4. An atmosphere of sufficient density, and consisting of the gases which are essential for vegetable and animal life. These are Oxygen, Carbonic-acid gas, Aqueous vapour, Nitrogen, and Ammonia. These must all be present in suitable proportions.

5. Alternations of day and night.

Small Range of Temperature required for

Growth and Development

Vital phenomena for the most part occur between the

temperatures of freezing water and 104° Fahr., and this is supposed to be due mainly to the properties of nitrogen and its compounds, which between these temperatures only can maintain those peculiarities which are essential to life—extreme sensitiveness and lability; facility of change as regards chemical combination and energy; and other properties which alone render nutrition, growth, and continual repair possible. A very small increase or decrease of temperature beyond these limits, if continued for any considerable time, would certainly destroy most existing forms of life, and would not improbably render any further development of life impossible except in some of its lowest forms.

As one example of the direct effects of increased temperature, we may adduce the coagulation of albumen. This substance is one of the proteids, and plays an important part in the vital phenomena of both plants and animals, and its fluidity and power of easy combination and change of form are lost by any degree of coagulation which takes place at about 160° Fahr.

The extreme importance to all the higher organisms of a moderate temperature is strikingly shown by the complex and successful arrangements for maintaining a uniform degree of heat in the interior of the body. The normal blood-heat in a man is 98° Fahr., and this is constantly maintained within one or two degrees though the external temperature may be more than fifty degrees below the freezing-point. High temperatures upon the earth's surface do not range so far from the mean as do the low. In the greater part of the tropics the air-temperature seldom reaches 96° Fahr., though in arid districts and deserts, which occur chiefly along the margins of the northern and southern tropics, it not unfrequently surpasses 110° Fahr., and even occasionally rises to 115° or 120° in Australia and Central India. Yet with suitable food and moderate care the blood-temperature of a healthy man would not rise or fall more than one or at most two degrees. The great importance of this uniformity of temperature in all the vital organs is distinctly shown by the fact that when, during fevers, the temperature of the patient rises six degrees above the normal amount, his condition is critical, while an increase of seven or eight degrees is an almost certain indication of a fatal result. Even in the vegetable kingdom seeds will not germinate under a temperature of four or five degrees above the freezing-point.

Now this extreme sensibility to variations of internal temperature is quite intelligible when we consider the complexity and instability of protoplasm, and of all the proteids in the living organism, and how important it is that the processes of nutrition and growth, involving constant motion of fluids and incessant molecular decompositions and recombinations, should be effected with the greatest regularity. And though a few of the higher animals, including man, are so perfectly organised that they can adapt or protect themselves so as to be able to live under very extreme conditions as regards temperature, yet this is not the case with the great majority, or with the lower types, as evidenced by the almost complete absence of reptiles from the arctic regions.

It must also be remembered that extreme cold and extreme heat are nowhere perpetual. There is always some diversity of seasons, and there is no land animal which passes its whole life where the temperature never rises above the freezing point.

The Necessity of Solar Light

Whether the higher animals and man could have been developed upon the earth without solar light, even if all the other essential conditions were present, is doubtful. That, however, is not the point I am at present considering, but one that is much more fundamental. Without plant life land animals at all events could never have come into existence, because they have not the power of making protoplasm out of inorganic matter. The plant alone can take the carbon out of the small proportion of carbonic acid in the atmosphere, and with it, and the other necessary elements, as already described, build up those wonderful carbon compounds which are the very foundation of animal life. But it does this solely by the agency of solar light, and even uses a special portion of that light. Not only, therefore, is a sun needed to give light and heat, but it is quite possible that *any* sun would not answer the purpose. A sun is required whose light possesses those special rays which are effective for this operation, and as we know that the stars differ greatly in their spectra, and therefore in the nature of their light, all might not be able to effect this great transformation, which is one of the very first steps in rendering animal life possible on our earth, and therefore probably on all earths.

Water a First Essential of Organic Life

It is hardly necessary to point out the absolute necessity of water, since it actually constitutes a very large proportion of the material of every living organism, and about three-fourths of our own bodies. Water, therefore, must be present everywhere, in one form or another, on any globe where life is possible. Neither animal nor plant can exist without it. It must also be present in such quantity and so distributed as to be constantly available on every part of a globe where life is to be maintained; and it is equally necessary that it should have persisted in equal profusion throughout those enormous geological epochs during which life has been developing. We shall see later on how very special are the conditions that have secured this continuous distribution of water on our earth, and we shall also learn that this large amount of water, its wide distribution, and its arrangement with regard to the land-surface, is an essential factor in producing that limited range of temperature which, as we have seen, is a primary condition for the development and maintenance of life.

The Atmosphere must be of Sufficient Density and Composed of Suitable Gases

The atmosphere of any planet on which life can be developed must have several qualities which are unconnected with each other, and the coincidence of which may be a rare phenomenon in the universe. The first of these is a sufficient density, which is required for two purposes—as a storer of heat, and in order to supply the oxygen, carbonic acid, and aqueous vapour in sufficient quantities for the requirements of vegetable and animal life.

As a reservoir of heat and a regulator of temperature, a rather dense atmosphere is a first necessity, in co-operation with the large quantity and wide distribution of water referred to in the last section. The very different character of our south-west from our north-east winds is a good illustration of its power of distributing heat and moisture. This it does owing to the peculiar property it possesses of allowing the sun's rays to pass freely through it to the earth which it warms, but acting like a blanket in preventing the rapid escape of the non-luminous heat so produced. But the heat stored up during the day is given out at night, and thus secures a uniformity of temperature which would not otherwise exist. This effect is strikingly seen at high altitudes, where the temperature becomes lower and lower, till at a not very great elevation, even in the tropics, snow lies on the ground all the year round. This is almost wholly due to the rarity of the air, which, on that account, has not so much capacity for heat. It also allows the heat it acquires to radiate more freely than denser air, so that the nights are much colder. At about 18,000 feet high our atmosphere is exactly half its density at the sea-level. This is considerably higher than the usual snow-line, even under the equator, whence it follows that if our atmosphere was only half its present density it would render the earth unsuitable for the higher forms of animal life. It is not easy to say exactly what would be the result as regards climate; but it seems likely that, except perhaps in limited areas in the tropics, where conditions were very favourable, the whole land-surface would become buried in snow and ice. This appears inevitable, because evaporation from the oceans by direct sun-heat would be more rapid than now; but as the vapour rose in the rare atmosphere it would rapidly become frozen, and snow would fall almost perpetually, although it might not lie permanently on the ground in the equatorial lowlands. It appears certain, therefore, that with half our present bulk of atmosphere life would be hardly possible on the earth on account of lowered temperature alone. And as there would certainly be an added difficulty in the needful supply of oxygen to animals and carbonic acid to plants, it seems highly probable that a reduction of density of even one-fourth might be sufficient to render a large portion of the globe a snow and ice-clad waste, and the remainder liable to such extremes of climate that only low forms of life could have arisen and been permanently maintained.

The Gases of the Atmosphere

Coming now to consider the constituent gases of the atmosphere, there is reason to believe that they form a mixture as nicely balanced in regard to animal and vegetable life as are the density and the temperature. At a first view of the subject we might conclude that oxygen is the one great essential for animal life, and that all else is of little importance. But further consideration shows us that nitrogen, although merely a diluent of the oxygen as regards the respiration of animals, is of the first importance to plants, which obtain it from the ammonia formed in the atmosphere and carried down into the soil by the rain. Although there is only one part of ammonia to a million of air, yet upon this minute proportion the very existence of the animal world depends, because neither animals nor plants can assimilate the free nitrogen of the air into their tissues.

Another fundamentally important gas in the atmosphere is carbonic acid, which forms about four parts in ten thousand parts of air, and, as already stated, is the source from which plants build up the great bulk of their tissues, as well as those protoplasms and proteids so absolutely necessary as food for animals. An important fact to notice here is, that carbonic acid, so essential to plants, and to animals through plants, is yet a poison to animals. When present in much more than the normal quantity, as it often is in cities and in badly ventilated buildings, it becomes highly prejudicial to health; but this is believed to be partly due to the various corporeal emanations and other impurities associated with it. Pure carbonic acid gas to the amount of even one per cent. in otherwise pure air can, it is said, be breathed for a time without bad effects, but anything more than this proportion will soon produce suffocation. It is probable, therefore, that a very much smaller proportion than one per cent., if constantly present, would be dangerous to life; though no doubt, if this had always been the proportion, life might have been developed in adaptation to it. Considering, however, that this poisonous gas is largely given out by the higher animals as a product of respiration, it would evidently be dangerous to the permanence of life if the quantity forming a constant constituent of the atmosphere were much greater than it is.

Aqueous Vapour in the Atmosphere

This water-gas, although it occurs in the atmosphere in largely varying quantities, is yet, in two distinct ways, essential to organic life. It prevents the too rapid loss of moisture from the leaves of plants when exposed to the sun, and it is also absorbed by the upper surface of the leaf and by the young shoots, which thus obtain both water and minute quantities of ammonia when the supply by the roots is insufficient. But it is of even more vital importance in supplying the hydrogen which, when united with the nitrogen of the atmosphere by electrical discharges, produces the ammonia, which is the main source of all the proteids of the plant, which proteids are the very foundation of animal life.

From this brief statement of the purposes served by the various gases forming our atmosphere, we see that they are to some extent antagonistic, and that any considerable increase of one or the other would lead to results that might be injurious either directly or in their ultimate results. And as the elements which constitute the bulk of all living matter possess properties which render them alone suitable for the purpose, we may conclude that the proportions in which they exist in our atmosphere cannot be very widely departed from wherever organic forms are developed.

The Alternation of Day and Night

Although it is difficult to decide positively whether alternations of light and darkness at short intervals are absolutely essential for the development of the various higher forms of life, or whether a world in which light was constant might do as well, yet on the whole it seems probable that day and night are really important factors. All nature is full of rhythmic movements of infinitely varied kinds, degrees, and durations. All the motions and functions of living things are periodic; growth and repair, assimilation and waste, go on alternately. All our organs are subject to fatigue and require rest. All kinds of stimulus must be of short duration or injurious results follow. Hence the advantage of darkness, when the stimuli of light and heat are partially removed, and we welcome 'tired nature's sweet restorer, balmy sleep'— giving rest to all the senses and faculties of body and mind, and endowing us with renewed vigour for another period of activity and enjoyment of life.

Plants as well as animals are invigorated by this nightly repose; and all alike benefit by these longer periods of greater and less amounts of work caused by summer and winter, dry and wet seasons. It is a suggestive fact, that where the influence of heat and light is greatest—within the tropics— the days and nights are of equal length, giving equal periods of activity and rest. But in cold and Arctic regions where, during the short summer, light is nearly perpetual, and all the functions of life, in vegetation especially, go on with extreme rapidity, this is followed by the long rest of winter, with its short days and greatly lengthened periods of darkness.

Of course, all this is rather suggestion than proof. It is possible that in a world of perpetual day or in one of perpetual night, life *might* have been developed. But on the other hand, considering the great variety of physical conditions which are seen to be necessary for the development and preservation of life in its endless varieties, any prejudicial influences, however slight, might turn the scale, and prevent that harmonious and continuous evolution which we know *must* have occurred.

So far I have only considered the question of day and night as regards the presence or absence of light. But it is probably far more important in its heat aspect; and here its period becomes of great, perhaps vital, importance. With its present duration of twelve hours day and twelve night on the average, there is not time, even between the tropics, for the earth to become so excessively heated as to be inimical to life; while a considerable portion of the heat, stored up in the soil, the water, and the atmosphere, is given out at night, and thus prevents a too sudden and injurious contrast of heat and cold. If the day and night were each very much longer—say 50 or 100 hours—it is quite certain that, during a day of that duration, the heat would become so great as to be inimical, perhaps prohibitive, to most forms of life; while the absence of all sun-heat for an equally long period would result in a temperature far below the freezing point of water. It is doubtful whether any high forms of animal life could have arisen under such great and continual contrasts of temperature.

We will now proceed to point out the special features which, in our earth, have combined to bring about and to maintain the various and complex conditions we have seen to be essential for life as it exists around us.

CHAPTER XII

THE EARTH IN ITS RELATION TO THE DEVELOPMENT AND MAINTENANCE OF LIFE

The first circumstance to be considered in relation to the habitability of a planet is its distance from the sun. We know that the heating power of the sun upon our earth is ample for the development of life in an almost infinite variety of forms; and we have a large amount of evidence to show that, were it not for the equalising power of air and water, distributed as they are with us, the heat received from the sun would be sometimes too great and sometimes too little. In some parts of Africa, Australia, and India, the sandy soil becomes so hot that an egg can be cooked by placing it just below the surface. On the other hand, at an elevation of about 12,000 feet in lat. 40° it freezes every night, and throughout the day in all places sheltered from the sun. Now, both these temperatures are adverse to life, and if either of them persisted over a considerable portion of the earth, the development of life would have been impossible. But the heat derived from the sun is inversely as the square of the distance, so that at half the distance we should have four times as much heat, and at twice the distance only one-fourth of the heat. Even at two-thirds of the distance we should receive more than twice as much heat; and, considering the facts as to the extreme sensitiveness of protoplasm and the coagulation of albumen, it seems certain that we are situated in what has been termed the temperate zone of the solar system, and that we could not be removed far from our present position without endangering a considerable portion of the life now existing upon the earth, and in all probability rendering the actual development of life, through all its phases and gradations, impossible.

The Obliquity of the Ecliptic

The effect of the obliquity of the earth's equator to its path round the sun, upon which depend our varying seasons and the inequality of day and night throughout all the temperate zones, is very generally known. But it is not usually considered that this obliquity is of any great importance as regards the suitability of the earth for the development and maintenance of life; and it seems to have been passed over as an accident hardly worth notice, because almost any other obliquity or none at all would have been equally advantageous. But if we consider what the direction of the earth's axis might possibly have been, we shall find that it is really a matter of great importance from our present point of view.

Let us suppose, first, that the earth's axis was, like that of Uranus, almost exactly in the plane of its orbit or directed towards the sun. There can be little doubt that such a position would have rendered our world unfitted for the development of life. For the result would be the most tremendous contrasts of the seasons; at mid-winter, on one half the globe, arctic night and more than arctic cold would prevail; while on the other half there would be a midsummer of continuous day with a vertical sun and such an amount of heat as nowhere exists with us. At the two equinoxes the whole globe would enjoy equal day and night, all our present tropics and part of the sub-tropical zone having the sun at noon so near to the zenith as to have the essential of a tropical climate. But the change to about a month of constant sunshine or a month of continuous night would be so rapid, that it seems almost impossible that either vegetable or animal life would ever have developed under such terrible conditions.

The other extreme direction of the earth's axis, exactly at right angles to the plane of the orbit, would be much more favourable, but would still have its disadvantages. The whole surface from equator to poles would enjoy equal day and night, and every part would receive the same amount of sun-heat all the year round, so that there would be no change of seasons; but the heat received would vary with the latitude.

In our latitude the sun's altitude at noon all the year would be less than 40°, the same as now occurs at the equinoxes, and we might therefore have a perpetual spring as regards temperature. But the constancy of the heat in the equatorial and tropical regions and of cold towards the poles would lead to a more constant and more rapid circulation of air, and we should probably experience such continuous north-westerly winds as to render our climate always cold and probably very damp. Near the poles the sun would always be on, or close to, the horizon, and would give so little heat that the sea might be perpetually frozen and the land deeply snow-buried; and these conditions would probably extend into the temperate zone, and possibly so far south as to render life impossible in our latitudes, since whatever results arose would be due to permanent causes, and we know how powerful are snow and ice to extend their sway over adjacent areas if not counteracted by summer heat or warm moist winds. On the whole, therefore, it seems probable that this position of the earth's axis would result in a much smaller portion of its surface being capable of supporting a luxuriant and varied vegetable and animal life than is now the case; while the extreme uniformity of conditions everywhere present might be so antagonistic to the great law of rhythm that seems to pervade the universe, and be in other ways so unfavourable, that life-development would probably have taken quite a different course from that which it has taken.

Itappearsalmostcertain, therefore, thatsomeintermediate position of the axis would be the most favourable; and that which actually exists seems to combine the advantage of change of seasons with good climatical conditions over the largest possible area. We know that during the greater part of the epoch of life-development this area was much greater than at present, since a luxuriant vegetation of deciduous and evergreen trees and shrubs extended up to and within the Arctic Circle, leading to the formation of coal-beds both in palæozoic and tertiary times; the extremely favourable conditions for organic life which then prevailed over so large a portion of the globe's surface, and which persisted down to a comparatively recent epoch, lead to the conclusion that no more favourable degree of obliquity was possible than that which we actually possess. A short account of the evidence on this interesting subject will now be given.

Persistence of Mild Climates through Geologic Time

The whole of the geological evidence goes to show that in remote ages the climate of the earth was generally more uniform, though perhaps not warmer, than it is now, and this can be best explained by a slightly different distribution of sea and land, which allowed the warm waters of the tropical oceans to penetrate into various parts of the continents (which were then more broken up than they are now), and also to extend more freely into the Arctic regions. So soon as we go back into the tertiary period, we find indications of a warmer climate in the north temperate zone; and when we reach the middle of that period, we find abundant indications, both in plant and animal remains, of mild climates near to the Arctic Circle, or actually within it.

On the west coast of Greenland, in 70° N. lat., there are found abundance of fossil plants very beautifully preserved, among which are many different species of oaks, beeches, poplars, plane-trees, vines, walnuts, plums, chestnuts, sequoias, and numerous shrubs—137 species in all, indicating a vegetation such as now grows in the north temperate parts of America and Eastern Asia. And even further north, in Spitzbergen, in N. lat. 78° and 79°, a somewhat similar flora is found, not quite so varied, but with oaks, poplars, birches, planes, limes, hazels, pines, and many aquatic plants such as may now be found in West Norway and in Alaska, nearly twenty degrees further south.

Still more remote, in the Cretaceous period, fossil plants have been found in Greenland, consisting of ferns, cycads, conifers, and such trees and shrubs as poplars, sassafras, andromedas, magnolias, myrtles, and many others, similar in character and often identical in species with fossils of the same period found in Central Europe and the United States, indicating a widespread uniformity of climate, such as would be brought about by the great ocean-currents carrying the warm waters of the tropics into the Arctic seas.

Still further back, in the Jurassic period, we have proofs of a mild climate in East Siberia and at Andö in Norway just within the Arctic Circle, in numerous plant remains, and also remains of great reptiles allied to those found in the same strata in all parts of the world. Similar phenomena occur in the still earlier Triassic period; but we will pass on to the much more remote Carboniferous period, during which most of the great coal-beds of the world were formed from a luxuriant vegetation, consisting mostly of ferns, giant horse-tails, and primitive conifers. The luxuriance of these plants, which are often found beautifully preserved and in immense quantities, is supposed to indicate an atmosphere in which carbonic acid gas was much more abundant than now; and this is rendered probable by the small number and low type of terrestrial animals, consisting of a few insects and amphibia.

But the interesting point is, that true coal-beds, with similar fossils to those of our own coal-measures, are found at Spitzbergen and at Bear Island in East Siberia, both far within the Arctic Circle, again indicating a great uniformity of climate, and probably a denser and more vapour-laden atmosphere, which would act as a blanket over the earth and preserve the heat brought to the Arctic seas by the ocean currents from the warmer regions.

The still earlier Silurian rocks are also found abundantly in the Arctic regions, but their fossils are entirely of marine animals. Yet they show the same phenomena as regards climate, since the corals and cephalopodous mollusca found in the Arctic beds closely resemble those of all other parts of the earth.

Many other facts indicate that throughout the enormous periods required for the development of the varied forms of life upon the earth, the great phenomena of nature were but little different from those that prevail in our own times. The slow and gentle processes by which the various vegetable and animal remains were preserved are shown by the perfect state in which many of the fossils exist. Often trunks of trees, cycads, and tree-ferns are found standing erect, with their roots still imbedded in the soil they grew in. Large leaves of poplars, maples, oaks, and other trees are often preserved in as perfect a state as if gathered by a botanist and dried between paper for his herbarium, and the same is especially the case with the beautiful ferns of the Permian and Carboniferous periods. Throughout these and most other formations well-preserved ripple-marks are found in the solidified mud or sand of old seashores, differing in no respect from similar marks to be found on almost every coast to-day. Equally interesting are the marks of rain-drops preserved in the rocks of almost all ages. Sir Charles Lyell has given illustrations of recent impressions of rain-drops on the extensive mud-flats of Nova Scotia, and also an illustration of rain-drops on a slab of shale from the carboniferous formation of the same country; and the two are as much alike as the prints of two different showers a few days apart. The general size and form of the drops are almost identical, and imply a great similarity in the general atmospheric conditions.

We must not forget that this presence of rain throughout geological time implies, as we have seen in our last chapter, a constant and universal distribution of atmospheric dust. The two chief sources of this dust—the total quantity of which in the atmosphere must be enormous—are volcanoes and deserts, and we are therefore sure that these two great natural phenomena have always been present. Of volcanoes we have ample independent evidence in the presence of lavas and volcanic ashes, as well as actual stumps or cores of old volcanoes, through all geological formations; and we can have little doubt that deserts also were present, though perhaps not always so extensive as they are now. It is a very suggestive fact that these two phenomena, usually held to be blots on the fair face of nature, and even to be opposed to belief in a beneficent Creator, should now be proved to be really essential to the earth's habitability.

Notwithstanding this prevalence of warm and uniform conditions, there is also evidence of considerable changes of climate; and at two periods—in the Eocene and in the remote Permian—there are even indications of ice-action, so that some geologists believe that there were then actual glacial epochs. But it seems more probable that they imply only local glaciation, owing to there having been high land and other suitable conditions for the production of glaciers in certain areas.

The whole bearing of the geological evidence indicates the wonderful continuity of conditions favourable for life, and for the most part of climatal conditions more favourable than those now prevailing, since a larger extent of land towards the North Pole was available for an abundant vegetation, and in all probability for an equally abundant animal life. We know, too, that there was never any total break in life-development; no epoch of such lowering or raising of temperature as to destroy all life; no such general subsidence as to submerge the whole land-surface. Although the geological record is in parts very imperfect, yet it is, on the whole, wonderfully complete; and it presents to our view a continuous progress, from simple to complex, from lower to higher. Type after type becomes highly specialised in adaptation to local or climatal conditions, and then dies out, giving room for some other type to arise and be specialised in harmony with the changed conditions. The general character of the inorganic change appears to have been from more insular to more continental conditions, accompanied by a change from more uniform to less uniform climates, from an almost sub-tropical warmth and moisture, extending up to the Arctic Circle, to that diversity of tropical, temperate, and cold areas, capable of supporting the greatest possible variety in the forms of life, and which seems especially adapted to stimulate mankind to civilisation and social development by means of the necessary struggle against, and utilisation of, the various forces of nature.

Water, its Amount and Distribution on the Earth

Although it is generally known that the oceans occupy more than two-thirds of the whole surface of the globe, the enormous bulk of the water in proportion to the land that rises above its surface is hardly ever appreciated. But as this is a matter of the greatest importance, both as regards the geological history of the globe and the special subject we are here discussing, it will be necessary to enter into some details in regard to it.

According to the best recent estimates, the land area of the globe is 0.28 of the whole surface, and the water area 0.72. But the mean height of the land above the sea-level is found to be 2250 feet, while the mean depth of the seas and oceans is 13,860 feet; so that though the water area is two and a half times that of the land, the mean depth of the water is more than six times the mean height of the land. This is, of course, due to the fact that lowlands occupy most of the land-area, the plateaus and high mountains a comparatively small portion of it; while, though the greatest depths of the oceans about equal the greatest heights of the mountains, yet over enormous areas the oceans are deep enough to submerge all the mountains of Europe and temperate North America, except the extreme summits of one or two of them. Hence it follows that the bulk of the oceans, even omitting all the shallow seas, is more than thirteen times that of the land above sea-level; and if all the land-surface and ocean-floors were reduced to one level, that is, if the solid mass of the globe were a true oblate spheroid, the whole would be covered with water about two miles deep. The diagram here given will render this more intelligible and will serve to illustrate what follows.

DIAGRAM OF PROPORTIONATE MEAN HEIGHT OF LAND AND AREADEPTH OF OCEANS.

Land Area. .28 of area of Globe. Ocean Area .72 of area of Globe.

In this diagram the lengths of the sections representing land and ocean are proportionate to their areas, while the thickness of each is proportionate to their mean height and mean depth respectively. Hence the two sections are in correct proportion to their cubic contents.

A mere inspection of this diagram is sufficient to disprove the old idea, still held by a few geologists and by many biologists, that oceans and continents have repeatedly changed places during geological times, or that the great oceans have again and again been bridged over to facilitate the distribution of beetles or birds, reptiles or mammals. We must remember that although the diagram shows the continents and oceans as a whole, yet it also shows, with quite sufficient accuracy, the proportions of each of the great continents to the oceans which are adjacent to them. It must also be borne in mind that there can be no elevation on a large scale without a corresponding subsidence elsewhere; because if there were not a vast unsupported hollow would be left beneath the rising land or in some part adjacent to it.

Now, looking at the diagram and at a chart or globe, try to imagine the ocean-bottom rising gradually, to form a continent joining Africa with South America or with Australia (both of which are demanded by many biologists): it is clear that, while such an elevation was going on, either some continental land or some other part of the ocean-bed must sink to a corresponding amount. We shall then see, that if such changes of elevation on a continental scale have taken place again and again at different periods, it would have been almost impossible, on every occasion, to avoid a whole continent being submerged (or even all the continents) in order to equalise subsidence with elevation while new continents were being raised up from the abyssal depths of the ocean. We conclude, therefore, that with the exception of a comparatively narrow belt around the continents, which may be roughly indicated by the thousand fathom line of soundings, the great ocean depths are permanent features of the earth's surface. It is this stability of the general distribution of land and water that has secured the continuity of life upon the earth. Had the great oceanic basins, on the other hand, been unstable, changing places with the land at various periods of geological time, they would, almost certainly, again and again have swallowed up the land in their vast abysses, and have thus destroyed all the organic life of the world.

There are many confirmatory proofs of this view (which is now widely accepted by geologists and physicists), and a few of them may be briefly stated.

1. None of the continents present us with marine deposits of any one geological age and occupying a large part of the surface of each, as must have been the case had they ever been sunk deep beneath the ocean and again elevated; neither do

any of them contain extensive formations corresponding to the deep oceanic clays and oozes, which again they must have done had they been at any time raised up from the ocean depths.

2. All the continents present an almost complete and continuous series of rocks of *all* geological ages, and in each of the great geological periods there are found fresh water and estuarine deposits, and even old land-surfaces, demonstrating continuity of continental or insular conditions.

3. All the great oceans possess, scattered over them, a few or many islands termed 'oceanic,' and characterised by a volcanic or coralline structure, with no ancient stratified rocks in anyone of them; and in none of these is there found a single indigenous land mammal or amphibian. It is incredible that, if these oceans had ever contained extensive continents, and if these oceanic islands are—as even now they are often alleged to be—parts of these now submerged continents, not one fragment of any of the old stratified rocks, which characterise all existing continents, should remain to show their origin. In the Atlantic we find the Azores, Madeira, and St. Helena; in the Indian Ocean, Mauritius, Bourbon, and Kerguelen Island; in the Pacific, the Fiji, Samoan, Society, Sandwich, and Galapagos Islands, all without exception telling us the same tale, that they have been built up from the ocean depths by submarine volcanoes and coralline growths, but have never formed part of continental areas.

4. The contours of the floors of all the great oceans, now fairly well known through the soundings of exploring vessels and for submarine telegraph lines, also give confirmatory evidence that they have never been continental land. For if any part of them were a sunken continent, that part must have retained some impress of its origin. Some of the numerous mountain ranges which characterise *every* continent would have remained. We should find slopes of from 20° to 50° not uncommon, while valleys bordered by rocky precipices, as in Lake Lucerne and a hundred others, or isolated rock-walled mountains like Roraima, or ranges of precipices as in the Ghâts of India or the Fiords of Norway, would frequently be met with. But not a single feature of this kind has ever been found in the ocean abysses. Instead of these we have vast plains, which, if the water were removed, would appear almost exactly level, with no abrupt slopes anywhere. When we consider that deposits from the land never reach these remote ocean depths, and that there is no wave-action below a few hundred feet, these continental features once submerged would be indestructible; and their total absence is, therefore, itself a demonstration that none of the great oceans are on the sites of submerged continents.

How Ocean Depths were Produced

It is a very difficult problem to determine how the vast basins which are filled by the great oceans, especially that of the Pacific, were first produced. When the earth's surface was still in a molten state, it would necessarily take the form of a true oblate spheroid, with a compression at the poles due to its speed of rotation, which is supposed to have been very great. The crust formed by the gradual cooling of such a globe would be of the same general form, and, being thin, would easily be fractured or bent so as to accommodate itself to any unequal stresses from the interior. As the crust thickened and the whole mass slowly cooled and contracted, fissures and crumpling would occur, the former serving as outlets for volcanic activities whose results are found throughout all geological ages; the latter producing mountain chains in which the rocks are almost always curved, folded, or even thrust over each other, indicating the mighty forces due to the adjustments of a solid crust upon a shrinking fluid or semi-fluid interior.

But during this whole process there seem to be no forces at work that could lead to the production of such a feature as the Pacific, a vast depression covering nearly one-third of the whole surface of the globe. The Atlantic Ocean, being smaller and nearly opposite to the Pacific, but approximately of equal depth, may be looked upon as a complementary phenomenon which will be probably explained as a result of the same causes as the vaster cavity.

So far as I am aware, there is only one suggested cause of the formation of these great oceans that seems adequate; and as that cause is to some extent supported by quite independent astronomical evidence, and also directly bears upon the main subject of the present volume, it must be briefly considered.

A few years ago, Professor George Darwin, of Cambridge, arrived at a certain conclusion as to the origin of the moon, which is now comparatively well known by Sir Robert Ball's popular account of it in his small volume, *Time and Tide*. Briefly stated, it is as follows. The tides produce friction on the earth and very slowly increase the length of our day, and also cause the moon to recede further from us. The day is lengthened only by a small fraction of a second in a thousand years, and the moon is receding at an equally imperceptible rate. But as these forces are constant, and have always acted on the earth and moon, as we go back and back into the almost infinite past we come to a time when the rotation of the earth was so rapid that gravity at the equator could hardly retain its outer portion, which was spread out so that the form of the whole mass was something like a cheese with rounded edges. And about the same epoch the distance of the moon is found to have been so small that it was actually touching the earth. All this is the result of mathematical calculation from the known laws of gravitation and tidal effects; and as it is difficult to see how so large a body as the moon could have originated in any other way, it is supposed that at a still earlier period the moon and earth were one, and that the moon separated from the parent mass owing to centrifugal force generated by the earth's rapid rotation. Whether the earth was liquid or solid at this epoch, and exactly how the separation occurred, is not explained either by Professor Darwin or Sir Robert Ball; but it is a very suggestive fact that, quite recently, it has been shown, by means of the spectroscope, that double stars of short period *do* originate in this way from a single star, as already described in our sixth chapter; but in these cases it seems probable that the parent star is in a gaseous state.

These investigations of Professor G. Darwin have been made use of by the Rev. Osmond Fisher (in his very interesting and important work, *Physics of the Earth's Crust*) to account for the basins of the great oceans, the Pacific being the chasm left when the larger portion of the mass of the moon parted from the earth.

Adopting, as I do, the theory of the origin of the earth by meteoric accretion of solid matter, we must consider our planet as having been produced from one of those vast rings of meteorites which in great numbers still circulate round the sun, but which at the much earlier period now contemplated were both more numerous and much more extensive. Owing to irregularities of distribution in such a ring and through disturbance by other bodies, aggregations of various size would inevitably occur, and the largest of these would in time draw in to itself all the rest, and thus form a planet. During the early stages of this process the particles would be so small and would come together so gradually, that little heat would be produced, and there would result merely a loose aggregation of cold matter. But as the process went on and the mass of the incipient planet became considerable— perhaps half that of the earth—the rest of the ring would fall in with greater and greater velocity; and this, added to the gravitative compression of the growing mass might, when nearly its present size, have produced sufficient heat to liquefy the outer layers, while the central portion remained solid and to some extent incoherent, with probably large quantities of heavy gases in the interstices. When the amount of the meteoric accretions became so reduced as to be insufficient to keep up the heat to the melting-point, a crust would form, and might have reached about half or three-fourths of its present thickness when the moon became separated.

Let us now try to picture to ourselves what happened. We should have a globe somewhat larger than our earth is now, both because it then contained the material of the moon and also because it was hotter, revolving so rapidly as to be very greatly flattened at the poles; while the equatorial belt bulged out enormously, and would probably have separated in the form of a ring with a very slight increase of the time of rotation, which is supposed to have been about four hours. This globe would have a comparatively thin crust, beneath which there was molten rock to an unknown depth, perhaps a few hundreds, perhaps more than a thousand miles. At this time the attraction of the sun acting on the molten interior produced tides in it, causing the thin crust to rise and fall every two hours, but to so small an extent—only about a foot or so—as not necessarily to fracture it; but it is calculated that this slight rhythmic undulation coincided with the normal period of undulation due to such a large mass of heavy liquid, and so tended to increase the instability due to rapid rotation.

The bulk of the moon is about one-fiftieth part that of the earth, and an easy calculation shows us that, taking the area of the Pacific, Atlantic, and Indian Oceans combined as about two-thirds that of the globe, it would require a thickness (or depth) of about forty miles to furnish the material for the moon. We must, of course, assume that there were some inequalities in the thickness of the crust and in its comparative rigidity, so that when the critical moment came and the earth could no longer retain its equatorial protuberance against the centrifugal force due to rotation combined with the tidal undulations caused by the sun, instead of a continuous ring slowly detaching itself, the crust gave way in two or more great masses where it was weakest, and as the tidal wave passed under it and a quantity of the liquid substratum rose with it, the whole would break up and collect into a sub-globular mass a short distance from the earth, and continue revolving with it for some time at about the same rate as the surface had rotated. But as tidal action is always equal on opposite sides of a globe, there would be a similar disruption there, forming, it may be supposed, the Atlantic basin, which, as may be seen on a small globe, is almost exactly opposite a part of the Central Pacific. So soon as these two great masses had separated from the earth, the latter would gradually settle down into a state of equilibrium, and the molten matter of the interior, which would now fill the great oceanic basins up to a level of a few mile below the general surface would soon cool enough to form a thin crust. The larger portion of the nascent moon would gradually attract to itself the one or more smaller portions and form our satellite; and from that time tidal friction by both moon and sun would begin to operate and would gradually lengthen our day and, more rapidly, our month in the way explained in Sir Robert Ball's volume.

A very interesting point may now be referred to, because it seems confirmatory of this origin of the great ocean basins. In Mr. Osmond Fisher's work it is explained how the variations in the force of gravity, at numerous points all over the world, have been determined by observations with the pendulum, and also how these variations afford a measure of the thickness of the solid crust, which is of less specific gravity than the molten interior on which it rests. By this means a very interesting result was obtained. The observations on numerous oceanic islands proved that the sub-oceanic crust was considerably more dense than the crust under the continents, but also thinner, the result being to bring the average mass of the sub-oceanic crust and oceans to an equality with that of the continental crust, and this causes the whirling earth to be in a state of balance, or equilibrium. Now, both the thinness and the increased density of the crust seem to be well explained by this theory of the origin of the oceanic basins. The new crust would necessarily for a long time be thinner than the older portion, because formed so much later, but it would very soon become cool enough to allow the aqueous vapour of the atmosphere and that given off through fissures from the molten interior to collect in the ocean basins, which would thenceforth be cooled more rapidly and kept at a uniform temperature and also under a uniform pressure, and these conditions would lead to the steady and continuous increase of thickness, with a greater compactness of structure than in the continental areas. It is no doubt to this uniformity of conditions, with a lowering of the bottom temperature throughout the greater part of geological time, till it has become only a few degrees above the freezing-point, that we owe the remarkable persistence of the vast and deep ocean basins on which, as we have seen, the continuity of life on the earth has largely depended.

There is one other fact which lends some support to this theory of the origin of the ocean basins—their almost complete symmetry with regard to the equator. Both the Atlantic and Pacific basins extend to an equal distance north and south of the equator, an equality which could hardly have been produced by any cause not directly connected with the earth's rotation. The polar seas which are coterminous with the two great oceans are very much shallower, and cannot, therefore, be considered as forming part of the true oceanic basins.

Water as an Equaliser of Temperature

The importance of water in regulating the temperature of the earth is so great that, even if we had enough water on the land for all the wants of plants and animals, but had no great oceans, it is almost certain that the earth could not have produced and sustained the various forms of life which it now possesses.

The effect of the oceans is twofold. Owing to the great specific heat of water, that is, its property of absorbing heat slowly but to a large amount, and giving it out with equal slowness, the surface-waters of the oceans and seas are heated by the sun so that by the evening of a bright day they have become quite warm to a depth of several feet. But air has much less specific heat than water, a pound of water in cooling one degree being capable of warming four pounds of air one degree; but as air is 770 times as light as water, it follows that the heat from one cubic foot of water will warm more than 3000 cubic feet of air as much as it cools itself. Hence the enormous surface of the seas and oceans, the larger part of which is within the tropics, warms the whole of the lower and denser portions of the air, especially during the night, and this warmth is carried to all parts of the earth by the winds, and thus ameliorates the climate. Another quite distinct effect is due to the great ocean currents, like the Gulf Stream and the Japan Current, which carry the warm water of the tropics to temperate and arctic regions, and thus render many countries habitable which would otherwise suffer the rigour of an almost arctic winter. These currents are, however, directly due to the winds, and properly belong to the section on the atmosphere.

The other equalising action, due primarily to the great area of the seas and oceans, is a result of the vast evaporating surface from which the land derives almost all its water in the form of rain and rivers; and it is quite evident that if there were not sufficient water-surface to produce an ample supply of vapour for this purpose, arid districts would occupy more and more of the earth's surface. How much water-surface is necessary for life we do not know; but if the proportions of water and land-surfaces were reversed, it seems probable that the larger proportion of the earth might be uninhabitable. The vapour thus produced has also a very great effect in equalising temperature; but this also is a point which will come better under our next chapter on the atmosphere.

There are, however, some matters connected with the water-supply of the earth, and its relation to the development of life, that call for a few remarks here. What has determined the total quantity of water on the earth or on other planets does not appear to be known; but presumably it would depend, partially or wholly, on the mass of the planet being sufficient to enable it to retain by its gravitative force the oxygen and hydrogen of which water is composed. As the two gases are so easily combined to form water, but can only be separated under special conditions, its quantity would be dependent on the supply of hydrogen, which is but rarely found on the earth in a free state. The important fact, however, is, that we do possess so great a quantity of water, that if the whole surface of the globe was as regularly contoured as are the continents, and merely wrinkled with mountain chains, then the existing water would cover the whole globe nearly two miles deep, leaving only the tops of high mountains above its surface as rows of small islands, with a few larger islands formed by what are now the high plateaus of Tibet and the Southern Andes.

Now there seems no reason why this distribution of the water should not have occurred—in fact it seems probable that it would have occurred, had it not been for the fortunate coincidence of the formation of enormously deep ocean basins. So far as I am aware, no sufficient explanation of the formation of these basins has been given but that of Mr. Osmond Fisher, as here described, and that depends upon three unique circumstances: (1) the formation of a satellite at a very late period of the planet's development when there was already a rather thick crust; (2) the satellite being far larger in proportion to its primary than any other in the solar system; and (3) its having been produced by fission from its primary on account of extremely rapid rotation, combined with solar tides in its molten interior, and a rate of oscillation of that molten interior coinciding with the tidal period.

Whether this very remarkable theory of the origin of our moon is the true one, and if so, whether the explanation it seems to afford of the great oceanic basins is correct, I am not mathematician enough to judge. The tidal theory of the origin of the moon, as worked out mathematically by Professor G.H. Darwin, has been supported by Sir Robert Ball and accepted by many other astronomers; while the researches of the Rev. Osmond Fisher into the *Physics of the Earth's Crust*, together with his mathematical abilities and his practical work as a geologist, entitle his opinion on the question of the mode of origin of the ocean basins to the highest respect. And, as we have seen, the existence of these vast and deep ocean basins, produced by the agency of a series of events so remarkable as to be quite unique in the solar system, played an important part in rendering the earth fit for the development of the higher forms of animal life, while without them it seems not improbable that the conditions would have been such as to render any varied forms of terrestrial life hardly possible.

CHAPTER XIII

THE EARTH IN RELATION TO LIFE: ATMOSPHERIC CONDITIONS

We have seen in our tenth chapter that the physical basis of life—protoplasm—consists of the four elements, oxygen, nitrogen, hydrogen, and carbon, and that both plants and animals depend largely upon the free oxygen in the air to carry on their vital processes; while the carbonic acid and ammonia in the atmosphere seem to be absolutely essential to plants. Whether life could have arisen and have been highly developed with an atmosphere composed of different elements from ours it is, of course, impossible to say; but there are certain physical conditions which seem absolutely essential whatever may be the elements which compose it.

The first of these essentials is an atmosphere which shall be of such density at the surface of the planet, and of so great a bulk, as to be not too rare to fulfil its various functions at all altitudes where there is a considerable area of land. What determines the total quantity of gaseous matter on the surface of a planet will be, mainly, its mass, together with the average temperature of its surface.

The molecules of gases are in a state of rapid motion

in all directions, and the lighter gases have the most rapid motions. The average speed of the motion of the molecules has been roughly determined under varying conditions of pressure and temperature, and also the probable maximum and minimum rates, and from these data, and certain known facts as to planetary atmospheres, Mr. G. Johnstone Stoney, F.R.S., has calculated what gases will escape from the atmospheres of the earth and the other planets. He finds that all the gases which are constituents of air have such comparatively low molecular rates of motion that the force of gravity at the upper limits of the earth's atmosphere is amply sufficient to retain them; hence the stability in its composition. But there are two other gases, hydrogen and helium, which are both known to enter the atmosphere, but never accumulate so as to form any measurable portion of it, and these are found to have sufficient molecular motion to escape from it. With regard to hydrogen, if the earth were much larger and more massive than it is, so as to retain the hydrogen, disastrous consequences might ensue, because, whenever a sufficient quantity of this gas accumulated, it would form an explosive mixture with the oxygen of the atmosphere, and a flash of lightning or even the smallest flame would lead to explosions so violent and destructive as perhaps to render such a planet unsuited for the development of life. We appear, therefore, to be just at the major limit of mass to secure habitability, except in such planets as may have no continuous supply of free hydrogen.

Perhaps the most important mechanical functions of the atmosphere dependent on its density are: (1) the production of winds, which in many ways bring about an equalisation of temperature, and which also produce surface-currents on the ocean; and (2) the distribution of moisture over the earth by means of clouds which also have other important functions.

Winds depend primarily on the local distribution of heat in the air, especially on the great amount of heat constantly present in the equatorial zone, due to the sun being always nearly vertical at noon, and to its being similarly vertical at each tropic once a year, with a longer day, leading to even higher temperatures than at the equator, and producing also that continuous belt of arid lands or deserts which almost encircle the globe in the region of the tropics. Heated air being lighter, the colder air from the temperate zones continually flows towards it, lifting it up and causing it to flow over, as it were, to the north and south. But as the inflow comes from an area of less rapid to one of more rapid rotation, the course of the air is diverted, and produces the north-east and south-east trades; while the overflow from the equator going to an area of less rapid rotation, turns westward and produces the south-west winds so prevalent over the north Atlantic and the north temperate zone generally, and the north-west in the southern hemisphere.

It is outside the zone of the equable trade-winds, and in a region a few degrees on each side of the tropics, that destructive hurricanes and typhoons prevail. These are really enormous whirlwinds due to the intensely heated atmosphere over the arid regions already mentioned, causing an inrush of cool air from various directions, thus setting up a rotatory motion which increases in rapidity till equilibrium is restored. The hurricanes of the West Indies and Mauritius, and the typhoons of the Eastern seas, are thus caused. Some of these storms are so violent that no human structures can resist them, while the largest and most vigorous trees are torn to pieces or overturned by them. But if our atmosphere were much denser than it is, its increased weight would give it still greater destructive force; and if to this were added a somewhat greater amount of sun-heat—which might be due either to our greater proximity to the sun or to the sun's greater size or greater heat-intensity, these tempests might be so increased in frequency and violence as to render considerable portions of the earth uninhabitable.

The constant and equable trade-winds have a very important function in initiating those far-reaching ocean-currents which are of the greatest importance in equalising temperature. The well known Gulf Stream is to us the most important of these currents, because it plays the chief part in giving us the mild climate we enjoy in common with the whole of Western Europe, a mildness which is felt to a considerable distance within the Arctic Circle; and, in conjunction with the Japan current, which does the same for the whole of the temperate regions of the North Pacific, renders a large portion of the globe better adapted for life than it would be without these beneficial influences.

These equalising currents, however, are almost entirely due to the form and position of the continents, and especially to the fact that they are so situated as to leave vast expanses of ocean along the equatorial zone, and extending north and south to the arctic and antarctic regions. If with the same amount of land the continents had been so grouped as to occupy a considerable portion of the equatorial oceans—such as would have been the case had Africa been turned so as to join South America, and Asia been brought to the south-east so as to take the place of part of the equatorial Pacific, then the great ocean-currents could have been but feeble or have hardly existed. Without these currents much of the north and south temperate lands would have been buried in ice, while the largest portion of the continents would have been so intensely heated as perhaps to be unsuited for the development of the higher forms of animal life, since we have shown (in chapters X. and XI.) how delicate is the balance and how narrow the limits of temperature which are required.

There seems to be no reason whatever why some such

distribution of the sea and land should not have existed, had it not been for the admittedly exceptional conditions which led to the production of our satellite, thus necessarily forming vast chasms along the region of the equator where centrifugal force as well as the internal solar tides were most powerful, and where the thin crust was thus compelled to give way. And as the highest authorities declare that there are no indications of such an origin of satellites in the case of any other planet, the whole series of conditions favourable to life on the earth become all the more remarkable.

Clouds, their Importance and their Causes

Few persons have any adequate conception of the real nature of clouds and of the important part they take in rendering our world a habitable and an enjoyable one.

On the average, the rainfall over the oceans is much less than over the land, the whole region of the trade-winds having usually a cloudless sky and very little rain; but in the intervening belt of calms, near to the equator, a cloudy sky and heavy rains are frequent. This arises from the fact that the warm, moist air over the ocean is raised upwards, by the cold and heavy air from north and south, into a cooler region where it cannot hold so much aqueous vapour, which is there condensed and falls as rain. Generally, wherever the winds blow over extensive areas of water on to the land, especially if there are mountains or elevated plateaus which cause the moisture-laden air to rise to heights where the temperature is lower, clouds are formed and more or less rain falls. But if the land is of an arid nature and much heated by the sun, the air becomes capable of holding still more aqueous vapour, and even dense rain-clouds disperse without producing any rainfall. From these simple causes, with the large area of sea as compared with the land upon our earth, by far the larger portion of the surface is well supplied with rain, which, falling most abundantly in the elevated and therefore cooler regions, percolates the soil, and gives rise to those innumerable springs and rivulets which moisten and beautify the earth, and which, uniting together, form streams and rivers, which return to the seas and oceans whence they were originally derived.

Clouds and Rain Depend upon Atmospheric Dust

The beautiful system of aqueous circulation by means of the atmosphere as sketched above was long thought to explain the whole process, and to require no further elucidation; but about a quarter of a century back a curious experiment was made which indicated that there was another factor in the process which had been entirely overlooked. If a small jet of steam is sent into two large glass receivers, one filled with ordinary air, the other with air which has been filtered by passing through a thick layer of cotton wool so as to keep back all particles of solid matter, the first vessel will be instantly filled with condensed cloudy-looking vapour, while in the other vessel the air and vapour will remain perfectly transparent and invisible. Another experiment was then made to imitate more nearly what occurs in nature. The two vessels were prepared as before, but a small quantity of water was placed in each vessel and allowed to evaporate till the air was nearly saturated with vapour, which remained invisible in both. Both vessels were then slightly cooled, when instantly a dense cloud was formed in that filled with unfiltered air, while the other remained quite clear. These experiments proved that the mere cooling of air below the dew point will not cause the aqueous vapour in it to condense into drops so as to form mist, fog, or cloud, unless small particles of solid or liquid matter are present to act as nuclei upon which condensation begins. The density of a cloud will therefore depend not only on the quantity of vapour in the air, but on the presence of an abundance of minute dust-particles on which condensation can begin.

That such dust exists everywhere in the air, even up to great heights, is not a supposition but a proved fact. By exposing glass plates covered with glycerine in different places and at different altitudes the number of these particles in each cubic foot of air has been determined; and it is found that not only are they present everywhere at low levels, but that there are a considerable number even at the tops of the highest mountains. These solid particles also act in another way. By radiation in the higher atmosphere they become very cold, and thus condense the vapour by contact, just as the points of grass-blades condense it to form dew.

When steam is escaping from an engine we see a mass of dense white vapour, a miniature cloud; and if we are near it in cold, damp weather, we feel little drops of rain produced from it. But on a fine, warm day it rises quickly and soon melts away, and entirely disappears. Exactly the same thing happens on a larger scale in nature. In fine weather we may have abundant clouds continually passing high overhead, but they never produce rain, because as the minute globules of water slowly fall towards the earth, the warm dry air again turns them into invisible vapour. Again, in fine weather, we often see a small cloud on a mountain top which remains there a considerable time, even though a brisk wind is blowing. The mountain top is colder than the surrounding air, and the invisible vapour becomes condensed into cloud by passing over it, but the moment these cloud particles are carried past the summit into the warmer and drier air they are again evaporated and disappear. On Table Mountain, near Cape Town, this phenomenon occurs on a large scale, and is termed the table-cloth, the mass of white fleecy cloud seeming to hang over the flat mountain top to some distance down where it remains for several months, while all around there is bright sunshine.

Another phenomenon that indicates the universal presence of dust to enormous heights in the atmosphere is the blue colour of the sky. This is caused by the presence of such excessively minute particles of dust through an enormous thickness of the higher atmosphere—probably up to a height of twenty or thirty miles, or more—that they reflect only the light of short wave-length from the blue end of the spectrum. This also has been proved by experiment. If a glass cylinder, several feet long, is filled with pure air from which all solid particles have been removed by filtering and passing over red-hot platinum wires, and a ray of electric light is passed through it, the interior, when viewed laterally, appears quite dark, the light passing through in a straight line and not illuminating the air. But if a little more air is passed through the filter so rapidly as to allow only the minutest particles of dust to enter with it, the vessel becomes gradually filled with a blue haze, which gradually deepens into a beautiful blue, comparable with that of the sky. If now some of the unfiltered air is admitted, the blue fades away into the ordinary tint of daylight.

Since it has been known that liquid oxygen is blue, many people have concluded that this explains the blue colour of the sky. But it has really nothing to do with the point at issue. The blue of the liquid oxygen becomes so excessively faint in the gas, further attenuated as it is by the colourless nitrogen, that it would have no perceptible colour in the whole thickness of our atmosphere. Again, if it had a perceptible blue tint we could not see it against the blackness of space behind it; but white objects seen through it, such as the moon and clouds, should all appear blue, which they do not do. The blue we see is from the whole sky, and is therefore reflected light; and as pure air is quite transparent, there must be solid or liquid particles so minute as to reflect blue light only. In the lower atmosphere the rain-producing particles are larger, and reflect all the rays, thus diluting the blue colour near the horizon, and, by refraction and reflection combined, producing the various beautiful hues of sunrise and sunset.

This production of exquisite colours by the dust in the atmosphere, though adding greatly to the enjoyment of life, cannot be considered essential to it; but there is another circumstance connected with atmospheric dust which, though little appreciated, might have effects which can hardly be calculated. If there were no dust in the atmosphere, the sky would appear black even at noon, except in the actual direction of the sun; and the stars would be visible in the day as well as at night. This would follow because air does not reflect light, and is not visible. We should therefore receive no light from the sky itself as we do now, and the north side of every hill, house, and other solid objects, would be totally dark, unless there were any surfaces in that direction to reflect the light. The surface of the ground at a little distance would be in sunshine, and this would be the only source of light wherever direct sunlight was cut off. To get a good amount of pleasant light in houses it would be necessary to have them built on nearly level ground, or on ground rising to the north, and with walls of glass all round and down to the floor line, to receive as much as possible of the reflected light from the ground. What effect this kind of light would have on vegetation it is difficult to say, but trees and shrubs would probably grow laterally towards the south, east, and west, so as to get as much direct sunshine as possible.

A more important result would be that, as sunshine would be perpetual during the day, so much evaporation would take place that the soil would become arid and almost bare in places that are now covered with vegetation, and plants like the cactuses of Arizona and the euphorbias of South Africa would occupy a large portion of the surface.

Returning now from this collateral subject of light and colour to the more important aspect of the question—the absence of cloud and rain—we have to consider what would happen, and in what way the enormous quantity of water which would be evaporated under continual sunshine would be returned to the earth.

The first and most obvious means would be by abnormally abundant dews, which would be deposited almost every night on every form of leafy vegetation. Not only would all grass and herbage, but all the outer leaves of shrubs and trees, condense so much moisture as to take the place of rain so far as the needs of such vegetation were concerned. But without arrangements for irrigation cultivation would be almost impossible, because the bare soil would become intensely heated during the day, and would retain so much of its heat through the night so as to prevent any dew forming upon it.

Some more effective mode, therefore, of returning the aqueous vapour of the atmosphere to the earth and ocean, would be required, and this, I believe, would be done by means of hills and mountains of sufficient height to become decidedly colder than the lowlands. The air from over the oceans would be constantly loaded with moisture, and whenever the winds blew on to the land the air would be carried up the slopes of the hills into the colder regions, and there be rapidly condensed upon the vegetation, and also on the bare earth and rocks of northern slopes, and wherever they cooled sufficiently during the afternoon or night to be below the temperature of the air. The quantity of vapour thus condensed would reduce the atmospheric pressure, which would lead to an inrush of air from below, bringing with it more vapour, and this might give rise to perpetual torrents, especially on northern and eastern slopes. But as the evaporation would be much greater than at the present time, owing to perpetual sunshine, so the water returned to the earth would be greater, and as it would not be so uniformly distributed over the land as it is now, the result would perhaps be that extensive mountain sides would become devastated by violent torrents, rendering permanent vegetation almost impossible; while other and more extensive areas, in the absence of rain, would become arid wastes that would support only the few peculiar types of vegetation that are characteristic of such regions.

Whether such conditions as here supposed would prevent the development of the higher forms of life it is impossible to say, but it is certain that they would be very unfavourable, and might have much more disastrous consequences than any we have here suggested. We can hardly suppose that, with winds and rock-formations at all like what they are now, any world could be wholly free from atmospheric dust. If, however, the atmosphere itself were much less dense than it is, say one-half, which might very easily have been the case, then the winds would have less carrying power, and at the elevations at which clouds are usually formed there would not be enough dust-particles to assist in their formation. Hence fogs close to the earth's surface would largely take the place of clouds floating far above it, and these would certainly be less favourable to human life and to that of many of the higher animals than existing conditions.

The world-wide distribution of atmospheric dust is a remarkable phenomenon. As the blue colour of the sky is universal, the whole of the higher atmosphere must be pervaded by myriads of ultra-microscopical particles, which, by reflecting the blue rays only, give us not only the azure vault of heaven, but in combination with the coarser dust of lower altitudes, diffused daylight, the grand forms and motions of the fleecy clouds, and the 'gentle rain from heaven' to refresh the parched earth and make it beautiful with foliage and flowers. Over every part of the vast Pacific Ocean, whose islands must produce a minimum of dust, the sky is always blue, and its thousand isles do not suffer for want of rain. Over the great forest-plain of the Amazon valley, where the production of dust must be very small, there is yet abundance of rain-clouds and of rain. This is due primarily to the two great natural sources of dust—the active volcanoes, together with the deserts and more arid regions of the world; and, in the second place, to the density and wonderful mobility of the atmosphere, which not only carries the finest dust-particles to an enormous height, but distributes them through its whole extent with such wonderful uniformity.

Every dust-particle is of course much heavier than air, and in a comparatively short time, if the atmosphere were still, would fall to the ground. Tyndall found that the air of a cellar under the Royal Institution in Albemarle Street, which had not been opened for several months, was so pure that the path of a beam of electric light sent through it was quite invisible. But careful experiments show that not only is the air in continual motion, but the motion is excessively irregular, being hardly ever quite horizontal, but upwards and downwards and in every intermediate direction, as well as in countless whirls and eddies; and this complexity of motion must extend to a vast height, probably to fifty miles or more, in order to provide a sufficient thickness of those minutest particles which produce the blue of the sky.

All this complexity of motion is due to the action of the sun in heating the surface of the earth, and the extreme irregularity of that surface both as regards contour and its capacity for heat-absorption. In one area we have sand or rock or bare clay, which, when exposed to bright sunshine, become scorching hot; in another area we have dense vegetation, which, owing to evaporation caused by the sunshine, remains comparatively cool, and also the still cooler surfaces of rivers and Alpine lakes. But if the air were much less dense than it is, these movements would be less energetic, while all the dust that was raised to any considerable height would, by its own weight, fall back again to the earth much more rapidly than it does now. There would thus be much less dust permanently in the atmosphere, and this would inevitably lead to diminished rainfall and, partially, to the other injurious effects already described.

Atmospheric Electricity

We have already seen that vegetable organisms obtain the chief part of the nitrogen in their tissues from ammonia produced in the atmosphere and carried into the earth by rain. This substance can only be thus produced by the agency of electrical discharges, or lightning, which cause the combination of the hydrogen in the aqueous vapour with the free nitrogen of the air. But clouds are important agents in the accumulation of electricity in sufficient amount to produce the violent discharges we know as lightning, and it is doubtful whether without them there would be any discharges through the atmosphere capable of decomposing the aqueous vapour in it. Not only are clouds beneficial in the production of rain, and also in moderating the intensity of continuous sun-heat, but they are also requisite for the formation of chemical compounds in vegetables which are of the highest importance to the whole animal kingdom. So far as we know, animal life could not exist on the earth's surface without this source of nitrogen, and therefore without clouds and lightning; and these, we have just seen, depend primarily on a due proportion of dust in the atmosphere.

But this due proportion of dust is mainly supplied by

volcanoes and deserts, and its distribution and constant presence in the air depends upon the density of the atmosphere. This again depends on two other factors: the force of gravity due to the mass of the planet, and the absolute quantity of the free gases constituting the atmosphere.

We thus find that the vast, invisible ocean of air in which we live, and which is so important to us that deprivation of it for a few minutes is destructive of life, produces also many other beneficial effects of which we usually take little account, except at times when storm or tempest, or excessive heat or cold, remind us how delicate is the balance of conditions on which our comfort, and even our lives, depend.

But the sketch I have here attempted to give of its varied functions shows us that it is really a most complex structure, a wonderful piece of machinery, as it were, which in its various component gases, its actions and reactions upon the water and the land, its production of electrical discharges, and its furnishing the elements from which the whole fabric of organic life is composed and perpetually renewed, may be truly considered to be the very source and foundation of life itself. This is seen, not only in the fact of our absolute dependence upon it every minute of our lives, but in the terrible effects produced by even a slight degree of impurity in this vital element. Yet it is among those nations that claim to be the most civilised, those that profess to be guided by a knowledge of the laws of nature, those that most glory in the advance of science, that we find the greatest apathy, the greatest recklessness, in continually rendering impure this all-important necessary of life, to such a degree that the health of the larger portion of their populations is injured and their vitality lowered, by conditions which compel them to breathe more or less foul and impure air for the greater part of their lives. The huge and ever-increasing cities, the vast manufacturing towns belching forth smoke and poisonous gases, with the crowded dwellings, where millions are forced to live under the most terrible insanitary conditions, are the witnesses to this criminal apathy, this incredible recklessness and inhumanity.

For the last fifty years and more the inevitable results of such conditions have been fully known; yet to this day nothing of importance *has* been done, nothing is being done. In this beautiful land there is ample space and a superabundance of pure air for every individual. Yet our wealthy and our learned classes, our rulers and law-makers, our religious teachers and our men of science, all alike devote their lives and energies to anything or everything but this. Yet *this* is the one great and primary essential of a people's health and well-being, to which *everything* should, for the time, be subordinate. Till this is done, and done thoroughly and completely, our civilisation is naught, our science is naught, our religion is naught, and our politics are less than naught—are utterly despicable; are below contempt.

It has been the consideration of our wonderful atmosphere in its various relations to human life, and to all life, which has compelled me to this cry for the children and for outraged humanity. Will no body of humane men and women band themselves together, and take no rest till this crying evil is abolished, and with it nine-tenths of all the other evils that now afflict us? Let *everything* give way to this. As in a war of conquest or aggression nothing is allowed to stand in the way of victory, and all private rights are subordinated to the alleged public weal, so, in this war against filth, disease, and misery let nothing stand in the way—neither private interests nor vested rights—and we shall certainly conquer. This is the gospel that should be preached, in season and out of season, till the nation listens and is convinced. Let this be our claim: Pure air and pure water for every inhabitant of the British Isles. Vote for no one who says 'It can't be done.' Vote only for those who declare 'It shall be done.' It may take five or ten or twenty years, but all petty ameliorations, all piecemeal reforms, must wait till this fundamental reform is effected. Then, when we have enabled our people to breathe pure air, and drink pure water, and live upon simple food, and work and play and rest under healthy conditions, they will be in a position to decide (for the first time) what other reforms are really needed.

Remember! We claim to be a people of high civilisation, of advanced science, of great humanity, of enormous wealth!

For very shame do not let us say 'We *cannot* arrange matters so that our people may all breathe unpolluted, unpoisoned air!'

CHAPTER XIV

THE EARTH IS THE ONLY HABITABLE PLANET IN THE SOLAR SYSTEM

Having shown in the last three chapters how numerous and how complex are the conditions which alone render life possible on our earth, how nicely balanced are opposing forces, and how curious and delicate are the means by which the essential combinations of the elements are brought about, it will be a comparatively easy task to show how totally unfitted are all the other planets either to develop or to preserve the higher forms of life, and, in most cases, any forms above the lowest and most rudimentary. In order to make this clear we will take the most important of the conditions in order, and see how the various planets fulfil them.

Mass of a Planet and its Atmosphere

The height and density of the atmosphere of a planet is important as regards life in several ways. On its density depends its power of carrying moisture; of holding a sufficient supply of dust-particles for the formation of clouds; of carrying ultra-microscopic particles to such a height and in such quantity as to diffuse the light of the sun by reflection from the whole sky; of raising waves in the ocean and thus aerating its waters, and of producing the ocean currents which so greatly equalise temperature. Now this density depends on two factors: the mass of the planet and the quantity of the atmospheric gases. But there is good reason to think that the latter depends directly upon the former, because it is only when a certain mass is attained that any of the lighter permanent gases can be held on the surface of a planet. Thus, according to Dr. G. Johnstone Stoney, who has specially studied this subject, the moon cannot retain even such a heavy gas as carbonic acid, or the still heavier carbon disulphide; while no particle of oxygen, nitrogen, or water-vapour can possibly remain on it, owing to the fact of its mass being only about one-eightieth that of the earth. It is believed that there are considerable quantities of gases in the stellar spaces, and probably also within the solar system, but perhaps in the liquid or solid form. In that state they might be attracted by any small mass such as the moon, but the heat of its surface when exposed to the solar rays would quickly restore them to the gaseous condition, when they would at once escape.

It is only when a planet attains a mass at least a quarter that of the earth that it is capable of retaining water-vapour, one of the most essential of the gases; but with so small a mass as this, its whole atmosphere would probably be so limited in amount and so rare at the planet's surface that it would be quite unable to fulfil the various purposes for which an atmosphere is required in order to support life. For their adequate fulfilment the mass of a planet cannot be much less than that of the earth. Here we come to one of those nice adjustments of which so many have been already pointed out. Dr. Johnstone Stoney arrives at the conclusion that hydrogen escapes from the earth. It is continually produced in small quantities by submarine volcanoes, by fissures in volcanic regions, from decaying vegetation, and from some other sources; yet, though sometimes found in minute quantities, it forms no regular constituent of our atmosphere.

The quantity of hydrogen combined with oxygen to form the mass of water in our vast and deep oceans is enormous. Yet if it had been only one-tenth more than it actually is, the present land-surface would have been almost all submerged. How the adjustments occurred so that there was exactly enough hydrogen to fill the vast ocean basins with water to such a depth as to leave enough land-surface for the ample development of vegetable and animal life, and yet not so much as to be injurious to climate, it is difficult to imagine. Yet the adjustment stares us in the face. First, we have a satellite unique in size as compared with its primary, and apparently in lateness of origin; then we have a mode of origin for that satellite said to be certainly unique in the solar system; as a consequence of this origin, it is believed, we have enormously deep ocean basins symmetrically placed with regard to the equator—an arrangement which is very important for ocean circulation; then we must have had the right quantity of hydrogen, obtained in some unknown way, which formed water enough to fill these chasms, so as to leave an ample area of dry land, but which one-tenth more water would have ingulfed; and, lastly, we have oxygen enough left to form an atmosphere of sufficient density for all the requirements of life. It could not be that the surplus hydrogen escaped when the water had been produced, because it escapes very slowly, and it combines so easily with free oxygen by means of even a spark, as to make it certain that *all* the available hydrogen was used up in the oceanic waters, and that the supply from the earth's interior has been since comparatively small in amount.

There is yet one more adjustment to be noticed. All the facts now referred to show that the earth's mass is sufficient to bring about the conditions favourable for life. But if our globe had been a little larger, and proportionately denser, in all probability no life would have been possible. Between a planet of 8000 and one of 9500 miles diameter is not a large difference, when compared with the enormous range of size of the other planets. Yet this slight increase in diameter would give two-thirds increase in bulk, and, with a corresponding increase of density due to the greater gravitative force, the mass would be about double what it is. But with double the mass the quantity of gases of all sorts attracted and retained by gravity would probably have been double; and in that case there would have been double the quantity of water produced, as no hydrogen could then escape. But the *surface* of the globe would only be one half greater than at present, in which case the water would have sufficed to cover the whole surface several miles deep.

Habitability of Other Planets

When we look to the other planets of our system we see everywhere illustrations of the relation of size and mass to habitability. The smaller planets, Mercury and Mars, have not sufficient mass to retain water-vapour, and, without it, they cannot be habitable. All the larger planets can have very little solid matter, as indicated by their very low density notwithstanding their enormous mass. There is, therefore, very good reason for the belief that the adaptability of a planet for a full development of life is *primarily* dependent, within very narrow limits, on its size and, more directly, on its mass. But if the earth owes its specially constituted atmosphere and its nicely adjusted quantity of water to such general causes as here indicated, and the same causes apply to the other planets of the solar system, then the only planet on which life can be possible is Venus. As, however, it may be urged that exceptional causes may have given other planets an equal advantage in the matter of air and water, we will briefly consider some of the other conditions which we have found to be essential in the case of the earth, but which it is almost impossible to conceive as existing, to the required extent, on any of the other planets of the solar system.

A Small and Definite Range of Temperature

We have already seen within what narrow limits the temperature on a planet's surface must be maintained in order to develop and support life. We have also seen how numerous and how delicate are the conditions, such as density of atmosphere, extent and permanence of oceans, and distribution of sea and land, which are requisite, even with us, in order to render possible the continuous preservation of a sufficiently uniform temperature. Slight alterations one way or another might render the earth almost uninhabitable, through its being liable to alternations of too great heat or excessive cold. How then can we suppose that any other of the planets, which have either very much more or very much less sun-heat than we receive, could, by any possible modification of conditions, be rendered capable of producing and supporting a full and varied life-development?

Mars receives less than half the amount of sun-heat per unit of surface that we do. And as it is almost certain that it contains no water (its polar snows being caused by carbonic acid or some other heavy gas) it follows that, although it may produce vegetable life of some low kinds, it must be quite unsuited for that of the higher animals. Its small size and mass, the latter only one-ninth that of the earth, may probably allow it to possess a very rare atmosphere of oxygen and nitrogen, if those gases exist there, and this lack of density would render it unable to retain during the night the very moderate amount of heat it might absorb during the day. This conclusion is supported by its low reflecting power, showing that it has hardly any clouds in its scanty atmosphere. During the greater part of the twenty-four hours, therefore, its surface-temperature would probably be much below the freezing point of water; and this, taken in conjunction with the total absence of aqueous vapour or liquid water, would add still further to its unsuitability for animal life.

In Venus the conditions are equally adverse in the other direction. It receives from the sun almost double the amount of heat that we receive, and this alone would render necessary some extraordinary combination of modifying agencies in order to reduce and render uniform the excessively high temperature. But it is now known that Venus has one peculiarity which is in itself almost prohibitive of animal life, and probably of even the lowest forms of vegetable life. This peculiarity is, that through tidal action caused by the sun, its day has been made to coincide with its year, or, more properly, that it rotates on its axis in the same time that it revolves round the sun. Hence it always presents the same face to the sun; and while one half has a perpetual day, the other half has perpetual night, with perpetual twilight through refraction in a narrow belt adjoining the illuminated half. But the side that never receives the direct rays of the sun must be intensely cold, approximating, in the central portions, to the zero of temperature, while the half exposed to perpetual sunshine of double intensity to ours must almost certainly rise to a temperature far too great for the existence of protoplasm, and probably, therefore, of any form of animal life.

Venus appears to have a dense atmosphere, and its brilliancy suggests that we see the upper surface of a cloud-canopy, and this would no doubt greatly reduce the excessive solar heat. Its mass, being a little more than three-fourths that of the earth, would enable it to retain the same gases as we possess. But under the extraordinary conditions that prevail on the surface of this planet, it is hardly possible that the temperature of the illuminated side can be preserved in a sufficient state of uniformity for the development of life in any of its higher forms.

Mercury possesses the same peculiarity of keeping one face always towards the sun, and as it is so much smaller and so much nearer the sun its contrasts of heat and cold must be still more excessive, and we need hardly discuss the possibility of this planet being habitable. Its mass being only one-thirtieth that of the earth, water-vapour will certainly escape from it, and, most probably, nitrogen and oxygen also, so that it can possess very little atmosphere; and this is indicated by its low reflecting power, no less than 83 per cent. of the sun's light being absorbed, and only 17 per cent. reflected, whereas clouds reflect 72 per cent. This planet is therefore intensely heated on one side and frozen on the other; it has no water and hardly any atmosphere, and is therefore, from every point of view, totally unfitted for supporting living organisms.

Even if it is supposed that, in the case of Venus, its perpetual cloud-canopy may keep down the surface temperature within the limits necessary for animal life, the extraordinary turmoil in its atmosphere caused by the excessively contrasted temperatures of its dark and light hemispheres must be extremely inimical to life, if not absolutely prohibitive of it. For on the greater part of the hemisphere that never receives a ray of light or heat from the sun all the water and aqueous vapour must be turned into ice or snow, and it seems almost impossible that the air itself can escape congelation. It could only do so by a very rapid circulation of the whole atmosphere, and this would certainly be produced by the enormous and permanent difference of temperature between the two hemispheres. Indications of refraction by a dense atmosphere are visible during the planet's transit over the sun's disc, and also when it is in conjunction with the sun, and the refraction is so great that Venus is believed to have an atmosphere much higher than ours. But during the rapid circulation of such an atmosphere, heated on one half the planet and cooled on the other, most of the aqueous vapour must be taken out of it on the dark side as fast as it is produced on the heated side, though sufficient may remain to produce a canopy of very lofty clouds analogous to our cirri. The occasional visibility of the dark side of Venus may be caused by an electrical glow due to the friction of the perpetually overflowing and inflowing atmosphere, this being increased by reflection from a vast surface of perpetual snow. If we consider all the exceptional features of this planet, it appears certain that the conditions as regards climate cannot now be such as to maintain a temperature within the narrow limits essential for life, while there is little probability that at any earlier period it can have possessed and maintained the necessary stability during the long epochs which are requisite for its development.

Before considering the condition of the larger planets, it

will be well to refer to an argument which has been supposed to minimise the difficulties already stated as to those planets which approach nearest to the earth in size and distance from the sun.

The Argument from Extreme Conditions, on the Earth

In reply to the evidence showing how nice are the adaptations required for life-development, it is often objected that life does *now* exist under very extreme conditions— under tropic heat and arctic snows; in the burnt-up desert as well as in the moist tropical forest; in the air as well as in the water; on lofty mountains as well as on the level lowlands. This is no doubt true, but it does not prove that life could have been developed in a world where any of these extremes of climate characterised the whole surface. The deserts are inhabited because there are oases where water is attainable, as well as in the surrounding fertile areas. The arctic regions are inhabited because there is a summer, and during that summer there is vegetation. If the surface of the ground were always frozen, there would be no vegetation and no animal life.

The late Mr. R.A. Proctor put this argument of the diversity of conditions under which life actually does exist on the earth as well probably as it can be put. He says: 'When we consider the various conditions under which life is found to prevail, that no difference of climatic relations, or of elevation, of land, or of air, or of water, of soil in land, of freshness or saltness in water, of density in air, appears (so far as our researches have extended) to render life impossible, we are compelled to infer that the power of supporting life is a quality which has an exceedingly wide range in nature.'

This is true, but with certain reservations. The only species of animal which does really exist under the most varied conditions of climate is man, and he does so because his intellect renders him to some extent the ruler of nature. None of the lower animals have such a wide range, and the diversity of conditions is not really so great as it appears to be. The strict limits are nowhere permanently overpassed, and there is always the change from winter to summer, and the possibility of migration to less inhospitable areas.

The Great Planets all Uninhabitable

Having already shown that the condition of Mars, both as regards water, atmosphere, and temperature, is quite unfitted to maintain life, a view in which both general principles and telescopic examination perfectly agree, we may pass on to the outer planets, which, however, have long been given up as adapted for life even by the most ardent advocates for 'life in other worlds.' Their remoteness from the sun—even Jupiter being five times as far as the earth, and therefore receiving only one twenty-fifth of the light and heat that we receive per unit of surface—renders it almost impossible, even if other conditions were favourable, that they should possess surface-temperatures adequate to the necessities of organic life. But their very low densities, combined with very large size, renders it certain that they none of them have a solidified surface, or even the elements from which such a surface could be formed.

It is supposed that Jupiter and Saturn, as well as Uranus and Neptune, retain a considerable amount of internal heat, but certainly not sufficient to keep the metallic and other elements of which the sun and earth consist in a state of vapour, for if so they would be planetary stars and would shine by their own light. And if any considerable portion of their bulk consisted of these elements, whether in a solid or a liquid state, their densities would necessarily be much greater than that of the earth instead of very much less—Jupiter is under one-fourth the density of the earth, Saturn under an eighth, while Uranus and Neptune are of intermediate densities, though much less in bulk even than Saturn.

It thus appears that the solar system consists of two groups of planets which differ widely from each other. The outer group of four very large planets are almost wholly gaseous, and probably consist of the permanent gases—those which can only be liquefied or solidified at a very low temperature. In no other way can their small density combined with enormous bulk be accounted for.

The inner group also of four planets are totally unlike the preceding. They are all of small size, the earth being the largest. They are all of a density roughly proportionate to their bulk. The earth is both the largest and the densest of the group; not only is it situated at that distance from the sun which, through solar heat alone, allows water to remain in the liquid state over almost the whole of its surface, but it possesses numerous characteristics which secure a very equable temperature, and which have secured to it very nearly the same temperature during those enormous geological periods in which terrestrial life has existed. We have already shown that no other planet possesses these characteristics now, and it is almost equally certain that they never have possessed them in the past, and never will possess them in the future.

A Last Argument for Habitability of the Planets

Although it has been admitted by the late Mr. Proctor and some other astronomers that most of the planets are not *now* habitable, yet, it is often urged, they may have been so in the past or may become so in the future. Some are now too hot, others are now too cold; some have now no water, others have too much; but all go through their appointed series of stages, and during some of these stages life may be or may have been possible. This argument, although vague, will appeal to some readers, and it may, therefore, be necessary to reply to it. This is the more necessary as it is still made use of by astronomers. In a criticism of my article in *The Fortnightly Review*, M. Camille Flammarion, of the Paris Observatory, dramatically remarks: 'Yes, life is universal, and eternal, for time is one of its factors. Yesterday the moon, to-day the earth, to-morrow Jupiter. In space there are both cradles and tombs.'

It is thus suggested that the moon was once inhabited and that Jupiter will be inhabited in some remote future; but no attempt is made to deal with the essential physical conditions of these very diverse objects, rendering them not only *now*, but always, unfitted to develop and to maintain terrestrial or aerial life. This vague supposition—it can hardly be termed an argument—as regards past or future adaptability for life, of all the planets and some of the satellites in the solar system, is, however, rendered invalid by an equally general objection to which its upholders appear never to have given a moment's consideration; and as it is an objection which still further enforces the view as to the unique position of the earth in the solar system, it will be well to submit it to the judgment of our readers.

Limitation of the Sun's Heat

It is well known that there is, and has been for nearly half a century, a profound difference of opinion between geologists and physicists as to the actual or possible duration in years of life upon the earth. The geologists, being greatly impressed with the vast results produced by the slow processes of the wearing away of the rocks and the deposit of the material in seas or lakes, to be again upheaved to form dry land, and to be again carved out by rain and wind, by heat and cold, by snow and ice, into hills and valleys and grand mountain ranges; and further, by the fact that the highest mountains in every part of the globe very often exhibit on their loftiest summits stratified rocks which contain marine organisms, and were therefore originally laid down beneath the sea; and, yet again, by the fact that the loftiest mountains are often the most recent, and that these grand features of the earth's surface are but the latest examples of the action of forces that have been at work throughout all geological time—studying all their lives the detailed evidences of all these changes, have come to the conclusion that they imply enormous periods only to be measured by scores or hundreds of millions of years.

And the collateral study of fossil remains in the long series of rock-formations enforces this view. In the whole epoch of human history, and far back into prehistoric times during which man existed on the earth, although several animals have become extinct, yet there is no proof that any new one has been developed. But this human era, so far as yet known, going back certainly to the glacial epoch and almost certainly to pre-glacial times, cannot be estimated at less than a million, some think even several million years; and as there have certainly been some considerable alterations of level, excavation of valleys, deposits of great beds of gravel, and other superficial changes during this period, some kind of a scale of measurement of geological time has been obtained, by comparison with the very minute changes that have occurred during the historical period. This scale is admittedly a very imperfect one, but it is better than none at all; and it is by comparing these small changes with the far greater ones which have occurred during every successive step backward in geological history that these estimates of geological time have been arrived at. They are also supported by the palæontologists, to whom the vast panorama of successive forms of life is an ever-present reality. Directly they pass into the latest stage of the Tertiary period—the Pliocene of Sir Charles Lyell—all over the world new forms of life appear which are evidently the forerunners of many of our still existing species; and as they go a little further back, into the Miocene, there are indications of a warmer climate in Europe, and large numbers of mammals resembling those which now inhabit the tropics, but of quite distinct species and often of distinct genera and families. And here, though we have only reached to about the middle of the Tertiary period, the changes in the forms of life, in the climate, and in the land-surfaces are so great when compared with the very minute changes during the human epoch, as to require us to multiply the time elapsed many times over. Yet the whole of the Tertiary period, during which *all* the great groups of the higher animals were developed from a comparatively few generalised ancestral forms, is yet the shortest by far of the three great geological periods—the Mesozoic or Secondary, having been much longer, with still vaster changes both in the earth's crust and in the forms of life; while the Palæozoic or Primary, which carries us back to the earliest forms of life as represented by fossilised remains, is always estimated by geologists to be at least as long as the other two combined, and probably very much longer.

From these various considerations most geologists who have made any estimates of geological time from the period of the earliest fossiliferous rocks, have arrived at the conclusion that about 200 millions of years are required. But from the variety of the forms of life at this early period it is concluded that a very much greater duration is needed for the whole epoch of life. Speaking of the varied marine fauna of the Cambrian period, the late Professor Ramsay says:—'In this earliest known varied life we find no evidence of its having lived near the beginning of the zoological series. In a broad sense, compared with what must have gone before, both biologically and physically, all the phenomena connected with this old period seem, to my mind, to be of quite a recent description; and the climates of seas and lands were of the very same kind as those the world enjoys at the present day.' And Professor Huxley held very similar views when he declared: 'If the very small differences which are observable between the crocodiles of the older Secondary formations and those of the present day furnish any sort of an approximation towards an estimate of the average rate of change among reptiles, it is almost appalling to reflect how far back in Palæozoic times we must go before we can hope to arrive at that common stock from which the crocodiles, lizards, *Ornithoscelida*, and *Plesiosauria*, which had attained so great a development in the Triassic epoch, must have been derived.'

Now, in opposition to these demands of the geologists, in which they are almost unanimous, the most celebrated physicists, after full consideration of all possible sources of the heat of the sun, and knowing the rate at which it is now expending heat, declare, with complete conviction, that our sun cannot have existed as a heat-giving body for so long a period, and they would therefore reduce the time during which life can possibly have existed on the earth to about one-fourth of that demanded by geologists. In one of his latest articles, Lord Kelvin says:—'Now we have irrefragable dynamics proving that the whole life of our sun as a luminary is a very moderate number of million years, probably less than 50 million, possibly between 50 and 100' (*Phil. Mag.*, vol. ii., Sixth Ser., p. 175, Aug. 1901). In my *Island Life* (chap. X.) I have myself given reasons for thinking that both the stratigraphical and biological changes may have gone on more quickly than has been supposed, and that geological time (meaning thereby the time during which the development of life upon the earth has been going on) may be reduced so as possibly to be brought within the maximum period allowed by physicists; but there will certainly be no time to spare, and any planets dependent on our sun whose period of habitability is either past or to come, cannot possibly have, or have had, sufficient time for the necessarily slow evolution of the higher life-forms. Again, all physicists hold that the sun is now cooling, and that its future life will be much less than its past. In a lecture at the Royal Institution (published in *Nature Series*, in 1889), Lord Kelvin says:—'It would, I think, be exceedingly rash to assume as probable anything more than twenty million years of the sun's light in the past history of the earth, or to reckon more than five or six million years of sunlight for time to come.'

These extracts serve to show that, unless either geologists

or physicists are very far from any approach to accuracy in their estimates of past or future age of the sun, there is very great difficulty in bringing them into harmony or in accounting for the actual facts of the geological history of the earth and of the whole course of life-development upon it. We are, therefore, again brought to the conclusion that there has been, and is, no time to spare; that the *whole* of the available past life-period of the sun has been utilised for life-development on the earth, and that the future will be not much more than may be needed for the completion of the grand drama of human history, and the development of the full possibilities of the mental and moral nature of man.

We have here, then, a very powerful argument, from a different point of view than any previously considered, for the conclusion that man's place in the solar system is altogether unique, and that no other planet either has developed or can develop such a full and complete life-series as that which the earth has actually developed. Even if the conditions had been more favourable than they are seen to be on other planets, Mercury, Venus, and Mars could not possibly have preserved equability of conditions long enough for life-development, since for unknown ages they must have been passing slowly towards their present wholly *unsuitable* conditions; while Jupiter and the planets beyond him, whose epoch of life-development is supposed to be in the remote future when they shall have slowly cooled down to habitability, will then be still more faintly illuminated and scantily warmed by a rapidly cooling sun, and may thus become, at the best, globes of solid ice. This is the teaching of science—of the best science of the twentieth century. Yet we find even astronomers who, more than any other exponents of science, should give heed to the teachings of the sister sciences to which they owe so much, indulging in such rhapsodies as the following:—'In our solar system, this little earth has not obtained any special privileges from Nature, and it is strange to wish to confine life within the circle of terrestrial chemistry.' And again: 'Infinity encompasses us on all sides, life asserts itself, universal and eternal, our existence is but a fleeting moment, the vibration of an atom in a ray of the sun, and our planet is but an island floating in the celestial archipelago, to which no thought will ever place any bounds.'

In place of such 'wild and whirling words,' I have endeavoured to state the sober conclusions of the best workers and thinkers as to the nature and origin of the world in which we live, and of the universe which on all sides surrounds us. I leave it to my readers to decide which is the more trustworthy guide.

CHAPTER XV

THE STARS—HAVE THEY PLANETARY SYSTEMS?

ARE THEY BENEFICIAL TO US?

Most of the writers on the Plurality of Worlds, from Fontenelle to Proctor, taking into consideration the enormous number of the stars and their apparent uselessness to our world, have assumed that many of them *must* have systems of planets circling round them, and that some of these planets, at all events, *must* possess inhabitants, some, perhaps, lower, but others no doubt higher than ourselves. One of our well-known modern astronomers, writing only ten years ago, adopts the same view. He says: 'The suns which we call stars were clearly not created for our benefit. They are of very little practical use to the earth's inhabitants. They give us very little light; an additional small satellite—one considerably smaller than the moon—would have been much more useful in this respect than the millions of stars revealed by the telescope. They must therefore have been formed for some other purpose…. We may therefore conclude, with a high degree of probability, that the stars—at least those with spectra of the solar type—form centres of planetary systems somewhat similar to our own.' The author then discusses the conditions necessary for life analogous to that of our earth, as regards temperature, rotation, mass, atmosphere, water, etc., and he is the only writer I have met with who has considered these conditions; but he touches on them very briefly, and he arrives at the conclusion that, in the case of the stars of solar type, it is probable that *one* planet, situated at a proper distance, would be fitted to support life. He estimates roughly that there are about ten million stars of this type, that is, closely resembling our sun, and that if only one in ten of these has a planet at the proper distance and properly constituted in other respects, there will be one million worlds fitted for the support of animal life. He therefore concludes that there are probably many stars having life-bearing planets revolving round them.

There are, however, many considerations not taken account of by this writer which tend to reduce very considerably the above estimate. It is now known that immense numbers of the stars of smaller magnitudes are nearer to us than are the majority of the stars of the first and second magnitudes, so that it is probable that these, as well as a considerable proportion of the very faint telescopic stars, are really of small dimensions. We have evidence that many of the brightest stars are much larger than our sun, but there are probably ten times as many that are much smaller. We have seen that the whole of the past light and heat-giving duration of our sun has, according to the best authorities, been only just sufficient for the development of life upon the earth. But the duration of a sun's heat-giving power will depend mainly upon its mass, together with its constituent elements. Suns which are much smaller than ours are, therefore, from that cause alone, unsuited to give adequate light and heat for a sufficient time, and with sufficient uniformity, for life-development on planets, even if they possess any at the right distance, and with the extensive series of nicely adjusted conditions which I have shown to be necessary.

Again, we must, probably, rule out as unfitted for life-development the whole region of the Milky Way, on account of the excessive forces there in action, as shown by the immense size of many of the stars, their enormous heat-giving power, the crowding of stars and nebulous matter, the great number of star-clusters, and, especially, because it is the region of 'new stars,' which imply collisions of masses of matter sufficiently large to become visible from the immense distance we are from them, but yet excessively small as compared with suns the duration of whose light is to be measured by millions of years. Hence the Milky Way is the theatre of extreme activity and motion; it is comparatively crowded with matter undergoing continual change, and is therefore not sufficiently stable for long periods to be at all likely to possess habitable worlds.

We must, therefore, limit our possible planetary systems suitable for life-development, to stars situated inside the circle of the Milky Way and far removed from it—that is, to those composing the solar cluster. These have been variously estimated to consist of a few hundred or many thousand stars—at all events to a very small number as compared with the 'hundreds of millions' in the whole stellar universe. But even here we find that only a portion are probably suitable. Professor Newcomb arrives at the conclusion—as have some other astronomers—that the stars in general have a much smaller mass in proportion to the light they give than our sun has; and, after an elaborate discussion, he finally concludes that the brighter stars are, on the average, much less dense than our sun. In all probability, therefore, they cannot give light and heat for so long a period, and as this period in the case of our sun has only been just sufficient, the number of suns of the solar type and of a sufficient mass may be very limited. Yet further, even among stars having a similar physical constitution to our sun and of an equal or greater mass, only a portion of their period of luminosity would be suitable for the support of planetary life. While they are in process of formation by accretions of solid or gaseous masses, they would be subject to such fluctuations of temperature, and to such catastrophic outbursts when any larger mass than usual was drawn towards them, that the whole of this period—perhaps by far the longest portion of their existence—must be left out of the account of planet-producing suns. Yet all these are to us stars of various degrees of brilliancy. It is almost certain that it is only when the growth of a sun is nearly completed, and its heat has attained a maximum, that the epoch of life-development is likely to begin upon any planets it may possess at the most suitable distance, and upon which all the requisite conditions should be present.

It may be said that there are great numbers of stars beyond our solar cluster and yet within the circle of the Milky Way, as well as others towards the poles of the Milky Way, which I have not here referred to. But of these regions very little is known, because it is impossible to tell whether stars in these directions are situated in the outer portion of the solar cluster or in the regions beyond it. Some astronomers appear to think that these regions may be nearly empty of stars, and I have endeavoured to represent what seems to be the general view on this very difficult subject in the two diagrams of the stellar universe at pp. <u>300</u>, <u>301</u>. The regions beyond our cluster and above or below the plane of the Milky Way are those where the small irresolvable nebulæ abound, and these may indicate that sun-formation is not yet active in those regions. The two charts of Nebulæ and Clusters at the end of the volume illustrate, and perhaps tend to support this view.

Double and Multiple Star Systems

We have already seen, in our sixth chapter, how rapid and extraordinary has been the discovery of what are termed spectroscopic binaries—pairs of stars so close together as to appear like a single star in the most powerful telescopes. The systematic search for such stars has only been carried on for a few years, yet so many have been already found, and their numbers are increasing so rapidly, as to quite startle astronomers. One of the chief workers in this field, Professor Campbell of the Lick Observatory, has stated his opinion that, as accuracy of measurement increases, these discoveries will go on till—'the star that is not a spectroscopic binary will prove to be the rare exception,'—and other astronomers of eminence have expressed similar views. But these close revolving star-systems are generally admitted to be out of the category of life-producing suns. The tidal disturbances mutually produced must be enormous, and this must be inimical to the development of planets, unless they were very close to each sun, and thus in the most unfavourable position for life.

We thus see that the result of the most recent researches

among the stars is entirely opposed to the old idea that the countless myriads of stars *all* had planets circulating round them, and that the ultimate purpose of their existence was, that they should be supporters of life, as our sun is the supporter of life upon the earth. So far is this from being the case, that vast numbers of stars have to be put aside as wholly unfitted for such a purpose; and when by successive eliminations of this nature we have reduced the numbers which may possibly be available to a few millions, or even to a few thousands, there comes the last startling discovery, that the entire host of stars is found to contain binary systems in such rapidly increasing numbers, as to lead some of the very first astronomers of the day to the conclusion that single stars may someday be found to be the rare exception! But this tremendous generalisation would, at one stroke, sweep away a large proportion of the stars which other successive disqualifications had spared, and thus leave our sun, which is certainly single, and perhaps two or three companion orbs, alone among the starry host as possible supporters of life on some one of the planets which circulate around them.

But we do not really *know* that any such suns exist. If they exist we do not *know* that they possess planets. If any do possess planets these may not be at the proper distance, or be of the proper mass, to render life possible. If these primary conditions should be fulfilled, and if there should possibly be not only one or two, but a dozen or more that so far fulfil the first few conditions which are essential, what probability is there that all the other conditions, all the other nice adaptations, all the delicate balance of opposing forces that we have found to prevail upon the earth, and whose combination here is due to exceptional conditions which exist in the case of no other known planet—should *all* be again combined in some of the possible planets of these possibly existing suns?

I submit that the probability is *now* all the other way. So long as we could assume that all the stars might be, in all essentials, like our sun, it seemed almost ludicrous to suppose that our sun alone should be in a position to support life. But when we find that enormous classes like the gaseous stars of small density, the solar stars while increasing in size and temperature, the stars which are much smaller than our sun, the nebulous stars, probably all the stars of the Milky Way, and lastly that enormous class of spectroscopic doubles—veritable Aaron's rods which threaten to swallow up all the rest—that all these are for various reasons unlikely to have attendant planets adapted to develop life, then the probabilities seem to be enormously against there being any considerable number of suns possessing attendant habitable earths. Just as the habitability of all the planets and larger satellites, once assumed as so extremely probable as to amount almost to a certainty, is now generally given up, so that in speculating on life in stellar systems Mr. Gore assumes that only *one* planet to each sun can be habitable; in like manner it may, and I believe will, turn out, that of all the myriad stars, the more we learn about them, the smaller and smaller will become the scanty residue which, with any probability, we can suppose to illuminate and vivify habitable earths. And when with this scanty probability we combine the still scantier probability that any such planet will possess simultaneously, and for a sufficiently long period, *all* the highly complex and delicately balanced conditions known to be essential for a full life-development, the conception that on this earth alone has such development been completed will not seem so wildly improbable a conjecture as it has hitherto been held to be.

Are the Stars Beneficial to Us?

When I suggested in my first publication on this subject that some emanations from the stars *might* be beneficial or injurious, and that a central position *might* be essential in order to render these emanations equable, one of my astronomical critics laughed the idea to scorn, and declared that 'we might wander into outer space without losing anything more serious than we lose when the night is cloudy and we cannot see the stars.' How my critic knows that this is so he does not tell us. He states it positively, with no qualification, as if it were an established fact. It may be as well to inquire, therefore, if there is any evidence bearing upon the point at issue.

Astronomers are so fully occupied with the vast number and variety of the phenomena presented by the stellar universe and the various difficult problems arising therefrom, that many lesser but still interesting inquiries have necessarily received little attention. Such a minor problem is the determination of how much heat or other active radiation we receive from the stars; yet a few observations have been made with results that are of considerable interest.

In the years 1900 and 1901 Mr. E.F. Nichols of the Yerkes Observatory made a series of experiments with a radiometer of special construction, to determine the heat emitted by certain stars. The result arrived at was, that Vega gave about $1/200000000$ of the heat of a candle at one metre distance, and Arcturus about 2.2 times as much.

In 1895 and 1896 Mr. G.M. Minchin made a series of experiments on the *Electrical Measurement of Starlight*, by means of a photo-electric cell of peculiar construction which is sensitive to the whole of the rays in the spectrum, and also to some of the ultra-red and ultra-violet rays. Combined with this was a very delicate electrometer. The telescope employed to concentrate the light was a reflector of two feet aperture. Mr. Minchin was assisted in the experiments by the late Professor G.F. Fitzgerald, F.R.S., of Trinity College,

Dublin, which may be considered guarantee of the accuracy of the observations. The following are the chief results obtained:—

	Source of Light.	Deflection in Millimetres	Light in Candles.	E. M. F. Volts.
1896	Candle at 10 feet distance	18.70		
	Betelgeuse (0.9 mag.)	12.80	0.685	0.026
	Aldebaran (1.1 mag.)	5.21	0.279	0.012
	Procyon (0.5 mag)	4.89	0.261	0.011
	Alpha Cygni (1.3 mag.)	4.90	0.262	0.011
	Polaris (2.1 mag.)	3.10	0.166	0.007
	1 volt.	432.00		
1895	Arcturus (0.3 mag.)	8.2	1.01	0.019
	Vega (0.1 mag)	11.5	1.42	0.026
	Candle at 10 feet	8.1		

N.B.—The standard candle shone directly on the cell, whereas the star's light was concentrated by a 2-foot mirror. The sensitive surface on which the light of the stars was concentrated was $^1/_{20}$ inch in diameter. We must therefore diminish the amount of candle light in this table in the proportion of the square of the diameter of the mirror (in $^1/_{20}$ of an inch) to one, equal to $^1/_{230400}$. If we make the necessary reduction in the case of Vega, and also equalise the distance at which the candle was placed, we find the following result:—

Observer.	Star.	Candle power at 10 ft.
Minchin.	Vega	$^1/_{162250}$
Nichols.	"	$^1/_{22000000}$

This enormous difference in the result is no doubt largely due to the fact that Mr. Nichols's apparatus measured heat alone, whereas Mr. Minchin's cell measured almost all the rays. And this is further shown by the fact that, whereas Mr. Nichols found Arcturus a red star, hotter than Vega a white one, Mr. Minchin, measuring also the light-giving and some of the chemical rays, found Vega considerably more energetic than Arcturus. These comparisons also suggest that other modes of measurement might give yet higher results, but it will no doubt be urged that such minute effects must necessarily be quite inoperative upon the organic world.

There are, however, some considerations which tend the other way. Mr. Minchin remarks on the unexpected fact that Betelgeuse produces more than double the electrical energy of Procyon, a much brighter star. This indicates that many of the stars of smaller visual magnitudes may give out a large amount of energy, and it is this energy, which we now know can take many strange and varied forms, that would be likely to influence organic life. And as to the quantity being too minute to have any effect, we know that the excessively minute amount of light from the very smallest telescopic stars produces such chemical changes on a photographic plate as to form distinct images, with comparatively small lenses or reflectors and with an exposure of two or three hours. And if it were not that the diffused light of the surrounding sky also acts upon the plate and blurs the faint images, much smaller stars could be photographed.

We know that not all the rays, but a portion only, are capable of producing these effects; we know also that there are many kinds of radiation from the stars, and probably some yet undiscovered comparable with the X rays and other new forms of radiation. We must also remember the endless variety and the extreme instability of the protoplasmic products in the living organism, many of which are perhaps as sensitive to special rays as is the photographic plate.

And we are not here limited to action for a few minutes or a few hours, but throughout the whole night and day, and continued whenever the sky is clear for months or years. Thus the cumulative effect of these very weak radiations may become important. It is probable that their action would be most influential on plants, and here we find all the conditions requisite for its accumulation and utilisation in the large amount of leaf-surface exposed to it. A large tree must present some hundreds of superficial feet of receptive surface, while even shrubs and herbs often have a leaf-area of greater superficial extent than the object-glasses of our largest telescopes. Some of the highly complex chemical processes that go on in plants may be helped by these radiations, and their action would be increased by the fact that, coming from every direction over the whole surface of the heavens, the rays from the stars would be able to reach and act upon every leaf of the densest masses of foliage. The large amount of growth that takes place at night may be in part due to this agency.

Of course all this is highly speculative; but I submit, in view of the fact that the light of the very faintest stars *does* produce distinct chemical changes, that even the very minute heat-effects are measureable, as well as the electro-motive forces caused by them; and further, that when we consider the millions, perhaps hundreds of millions of stars, all acting simultaneously on any organism which may be sensitive to them, the supposition that they do produce some effect, and possibly a very important effect, is not one to be summarily rejected as altogether absurd and not worth inquiring into.

It is not, however, these possible direct actions of the stars upon living organisms to which I attach much weight as regards our central position in the stellar universe. Further consideration of the subject has convinced me that the fundamental importance of that position is a physical one, as has already been suggested by Sir Norman Lockyer and some other astronomers. Briefly, the central position appears to be the only one where suns can be sufficiently stable and long-lived to be capable of maintaining the long process of life-development in any of the planets they may possess. This point will be further developed in the next (and concluding) chapter.

CHAPTER XVI

STABILITY OF THE STAR-SYSTEM: IMPORTANCE OF OUR CENTRAL POSITION: SUMMARY AND CONCLUSION

One of the greatest difficulties with regard to the vast system of stars around us is the question of its permanence and stability, if not absolutely and indefinitely, yet for periods sufficiently long to allow for the many millions of years that have certainly been required for our terrestrial life-development. This period, in the case of the earth, as I have sufficiently shown, has been characterised throughout by extreme uniformity, while a continuance of that uniformity for a few millions of years in the future is almost equally certain.

But our mathematical astronomers can find no indications of such stability of the stellar universe as a whole, if subject to the law of gravitation alone. In reply to some questions on this point, my friend Professor George Darwin writes as follows:—'A symmetrical annular system of bodies might revolve in a circle with or without a central body. Such a system would be unstable. If the bodies are of unequal masses and not symmetrically disposed, the break-up of the system would probably be more rapid than in the ideal case of symmetry.'

This would imply that the great annular system of the Milky Way is unstable. But if so, its existence at all is a greater mystery than ever. Although in detail its structure is very irregular, as a whole it is wonderfully symmetrical; and it seems quite impossible that its generally circular ring-like form can be the result of the chance aggregation of matter from any pre-existing different form. Star-clusters are equally unstable, or, rather, nothing is known or can be predicated about their stability or instability, according to Professors Newcomb and Darwin.

Mr. E. T. Whittaker (Secretary to the Royal Astronomical Society), to whom Professor G. Darwin sent my questions, writes:—'I doubt whether the principal phenomena of the stellar universe are consequences of the law of gravitation at all. I have been working myself at spiral nebulæ, and have got a first approximation to an explanation—but it is electro-dynamical and not gravitational. In fact, it may be questioned whether, for bodies of such tremendous extent as the Milky Way or nebulæ, the effect which we call gravitation is given by Newton's law; just as the ordinary formulæ of electrostatic attraction break down when we consider charges moving with very great velocities.'

Accepting these statements and opinions of two mathematicians who have given special attention to similar problems, we need not limit ourselves to the laws of gravitation as having determined the present form of the stellar universe; and this is the more important because we may thus escape from a conclusion which many astronomers seem to think inevitable, viz. that the observed proper motions of the stars cannot be explained by the gravitative forces of the system itself. In chapter VIII. of this work I have quoted Professor Newcomb's calculation as to the effect of gravitation in a universe of 100 million stars, each five times the mass of our sun, and spread over a sphere which it would take light 30,000 years to cross; then, a body falling from its outer limits to the centre could at the utmost acquire a velocity of twenty-five miles a second; and therefore, any body in any part of such a universe having a greater velocity would pass away into infinite space. Now, as several stars have, it is believed, much more than this velocity, it follows not only that they will inevitably escape from our universe, but that they do not belong to it, as their great velocity must have been acquired elsewhere. This seems to have been the idea of the astronomer who stated that, even at the very moderate speed of our sun, we should in five million years be deep in the actual stream of the Milky Way. To this I have already sufficiently replied; but I now wish to bring before my readers an excellent illustration of the importance of the late Professor Huxley's remark, that the results you got out of the 'mathematical mill' depend entirely on what you put into it.

In the *Philosophical Magazine* (January 1902) is a remarkable article by Lord Kelvin, in which he discusses the very same problem as that which Professor Newcomb had discussed at a much earlier date, but, starting from different assumptions, equally based on ascertained facts and probabilities deduced from them, brings out a very different result.

Lord Kelvin postulates a sphere of such a radius that a star at its confines would have a parallax of one-thousandth part of a second (0".001), equivalent to 3215 light-years. Uniformly distributed through this sphere there is matter equal in mass to 1000 million suns like ours. If this matter becomes subject to gravitation, it all begins to move at first with almost infinite slowness, especially near its centre; but nevertheless, in twenty-five million years many of these suns would have acquired velocities of from twelve to twenty miles a second, while some would have less and some probably more than seventy miles a second. Now such velocities as these agree generally with the measured velocities of the stars, hence Lord Kelvin thinks there may be as much matter as 1000 million suns within the above-named distance. He then states that if we suppose there to be 10,000 million suns within the same sphere, velocities would be produced very much greater than the known star-velocities; hence it is probable that there is very much less matter than 10,000 million times the sun's mass. He also states that if the matter were not uniformly distributed within the sphere, then, whatever was the irregularity, the acquired motions would be greater; again indicating that the 1000 million suns would be ample to produce the observed effects of stellar motion. He then calculates the average distance apart of each of the 1000 million stars, which he finds to be about 300 millions of millions of miles. Now the nearest star to our sun is about twenty-six million million of miles distant, and, as the evidence shows, is situated in the denser part of the solar cluster. This gives ample allowance for the comparative emptiness of the space between our cluster and the Milky Way, as well as of the whole region towards the poles of the Milky Way (as shown by the diagrams in chapter IV.), while the comparative density of extensive portions of the Galaxy itself may serve to make up the average.

Now, previous writers have come to a different conclusion from the same general line of argument, because they have started with different assumptions. Professor Newcomb, whose statement made some years back is usually followed, assumed 100 million stars each five times as large as our sun, equal to 500 million suns in all, and he distributed them equally throughout a sphere 30,000 light-years in diameter. Thus he has half the amount of matter assumed by Lord Kelvin, but nearly five times the extent, the result being that gravity could only produce a maximum speed of twenty-five miles a second; whereas on Lord Kelvin's assumption a maximum speed of seventy miles a second would be produced, or even more. By this latter calculation we find no insuperable difficulty in the speed of any of the stars being beyond the power of gravitation to produce, because the rates here given are the direct results of gravitation acting on bodies almost uniformly distributed through space. Irregular distribution, such as we see everywhere in the universe, might lead to both greater and less velocities; and if we further take account of collisions and near approaches of large masses resulting in explosive disruptions, we might have almost any amount of motion as the result, but as this motion would be produced by gravitation within the system, it could equally well be controlled by gravitation.

In order that my readers may better understand the calculations of Lord Kelvin, and also the general conclusions of astronomers as to the form and dimensions of the stellar universe, I have drawn two diagrams, one showing a plan on the central plane of the Milky Way, the other a section through its poles. Both are on the same scale, and they show the total diameter across the Milky Way as being 3600 light-years, or about half that postulated by Lord Kelvin for his hypothetical universe. I do this because the dimensions given by him are those which are sufficient to lead to motions near the centre such as the stars now possess in a minimum period of twenty-five million years after the initial arrangement he

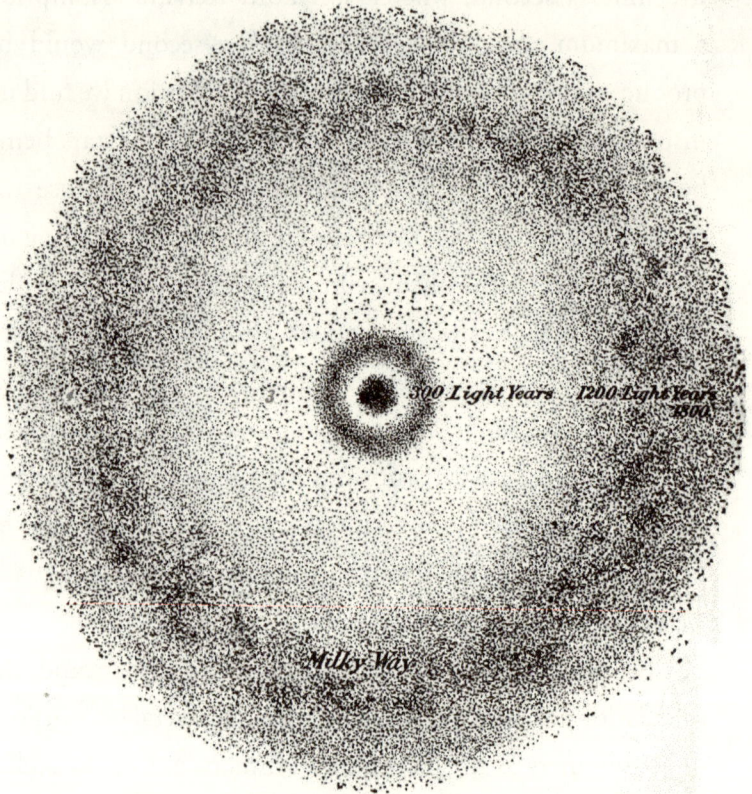

DIAGRAM OF STELLAR UNIVERSE (Plan).
1. Central part of Solar Cluster.
3. Outer limit of Solar Cluster.
2. Sun's Orbit (Black spot).   4. Milky Way.

supposes, at which later epoch which we are now supposed to have reached, the whole system would of course be greatly reduced in extent by aggregations towards and near the centre.

These dimensions also seem to accord sufficiently with the actual distances of stars as yet measured. The smallest parallax which has been determined with any certainty, according to Professor Newcomb's list, is that of Gamma Cassiopeiæ, which is one-hundredth of a second (0".01), while Lord Kelvin gives none smaller than 0".02, and these will all be included within the solar cluster as I have shown it.

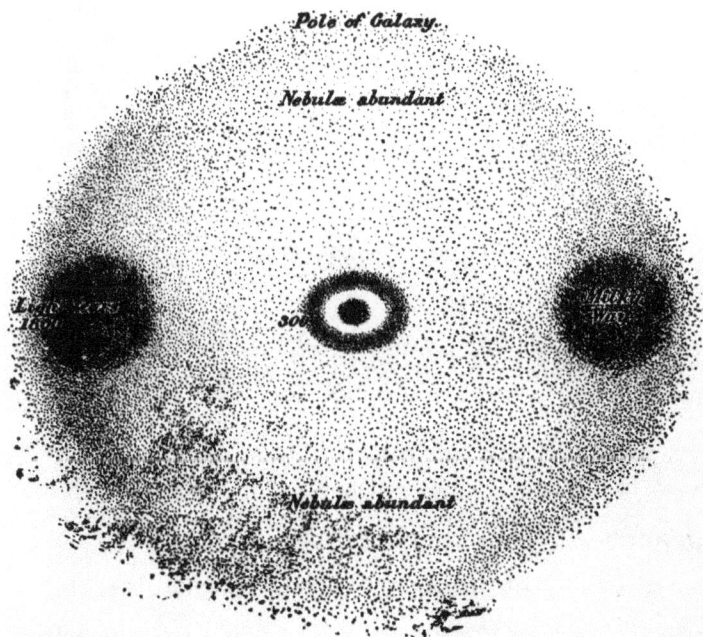

DIAGRAM OF STELLAR UNIVERSE (Section).
Section through Poles of Milky Way.

It must be clearly understood that these two illustrations are merely diagrams to show the main features of the stellar universe according to the best information available, with the proportionate dimensions of these features, so far as the facts of the distribution of the stars and the views of those astronomers who have paid most attention to the subject can be harmonised. Of course it is not suggested that the whole arrangement is so regular as here shown, but an attempt has been made by means of the dotted shading to represent the comparative densities of the different portions of space around us, and a few remarks on this point may be needed.

The solar cluster is shown very dense at the central portion, occupying one-tenth of its diameter, and it is near the outside of this dense centre that our sun is supposed to be situated. Beyond this there seems to be almost a vacuity, beyond which again is the outer portion of the cluster consisting of comparatively thinly scattered stars, thus forming a kind of ring-cluster, resembling in shape the beautiful ring-nebula in Lyra, as has been suggested by several astronomers. There is some direct evidence for this ring-form. Professor Newcomb in his recent book on *The Stars* gives a list of all stars of which the parallax is fairly well known. These are sixty-nine in number; and on arranging them in the order of the amount of their parallax, I find that no less than thirty-five of them have parallaxes between 0".1 and 0".4 of a second, thus showing that they constitute part of the dense central mass; while three others, from 0".4 to 0".75, indicate those which are our closest companions at the present time, but still at an enormous distance. Those which have parallaxes of less than the tenth and down to one-hundredth of a second are only thirty-one in all; but as they are spread over a sphere ten times the diameter, and therefore a thousand times the cubic content of the sphere containing those above one-tenth of a second, they ought to be immensely more numerous even if very much more thinly scattered. The interesting point, however, is, that till we get down to a parallax of 0".06, there are only three stars as yet measured, whereas those between 0".02 and 0".06, an equal range of parallax, are twenty-six in number, and as these are scattered in all directions they indicate an almost vacant space followed by a moderately dense outer ring.

In the enormous space between our cluster and the Milky Way, and also above and below its plane to the poles of the Galaxy, stars appear to be very thinly scattered, perhaps more densely in the plane of the Milky Way than above and below it where the irresolvable nebulæ are so numerous; and there may not improbably be an almost vacant space beyond our cluster for a considerable distance, as has been supposed, but this cannot be known till some means are discovered of measuring parallaxes of from one-hundredth to one five-hundredth of a second.

These diagrams also serve to indicate another point of considerable importance to the view here advocated. By placing the solar system towards the outer margin of the dense central portion of the solar cluster (which may very possibly include a large proportion of dark stars and thus be much more dense towards the centre than it appears to us), it may very well be supposed to revolve, with the other stars composing it, around the centre of gravity of the cluster, as the force of gravity towards that centre might be perhaps twenty or a hundred times greater than towards the very much less dense and more remote outer portions of the cluster. The sun, as indicated on the diagrams, is about thirty light-years from that centre, corresponding to a parallax of a little more than one-tenth of a second, and an actual distance of 190 millions of millions of miles, equal to about 70,000 times the distance of the sun from Neptune. Yet we see that this position is so little removed from the exact centre of the whole stellar universe, that if any beneficial influences are due to that central position in regard to the Galaxy, it will receive them perhaps to as full an extent as if situated at the actual centre. But if it is situated as here shown, there is no further difficulty as to its proper motion carrying it from one side to the other of the Milky Way in less time than has been required for the development of life upon the earth. And if the solar cluster is really sub-globular, and sufficiently condensed to serve as a centre of gravity for the whole of the stars of the cluster to revolve around, all the component stars which are not situated in the plane of its equator (and that of the Milky Way) must revolve obliquely at various angles up to an angle of 90°. These numerous diverging motions, together with the motions of the nearer stars outside the cluster, some of which may revolve round other centres of gravity made up largely of dark bodies, would perhaps sufficiently account for the apparent random motions of so many of the stars.

Uniform Heat-Supply due to Central Position

We now come to a point of the greatest interest as regards the problem we are investigating. We have seen how great is the difference in the estimates of geologists and those of physicists as to the time that has elapsed during the whole development of life. But the position we have now found for the sun, in the outer portion of the central star-cluster, may afford a clue to this problem. What we require is, some mode of keeping up the sun's heat during the enormous geological periods in which we have evidence of a wonderful uniformity in the earth's temperature, and therefore in the sun's heat-emission. The great central ring-cluster with its condensed central mass, which presumably has been forming for a much longer period than our sun has been giving heat to the earth, must during all this time have been exerting a powerful attraction on the diffused matter in the spaces around it, now apparently almost void as compared with what they may have been. Some scanty remnants of that matter we see in the numerous meteoric swarms which have been drawn into our system. A position towards the outside of this central aggregation of suns would evidently be very favourable for the growth by accretion of any considerable mass. The enormous distance apart of the outer components (the outer ring) of the cluster would allow a large amount of the inflowing meteoritic matter to escape them, and the larger suns situated near the surface of the inner dense cluster would draw to themselves the greater part of this matter. The various planets of our system were no doubt built up from a portion of the matter that flowed in near the plane of the ecliptic, but much of that which came from all other directions would be drawn towards the sun itself or to its neighbouring suns. Some of this would fall directly into it; other masses coming from different directions and colliding with each other would have their motion checked, and thus again fall into the sun; and so long as the matter falling in were not in too large masses, the slow additions to the sun's bulk and increase of its heat would be sufficiently gradual to be in no way prejudicial to a planet at the earth's distance.

The main point I wish to suggest here is, that by far the greater portion of the matter of the whole stellar universe has, either through gravitation or in combination with electrical forces, as suggested by Mr. Whittaker, become drawn together into the vast ring-formed system of the

Milky Way, which is, presumably, slowly revolving, and has thus been checked in its original inflow toward the centre of mass of the stellar universe. It has also probably drawn towards itself the adjacent portions of the scattered material in the spaces around it in all directions.

Had the vast mass of matter postulated by Lord Kelvin acquired no motion of revolution, but have fallen continuously towards the centre of mass, the motions developed when the more distant bodies approached that centre would have been extremely rapid; while, as they must have fallen in from every direction, they would have become more and more densely aggregated, and collisions of the most catastrophic nature would frequently have occurred, and this would have rendered the central portion of the universe the *least* stable and the *least* fitted to develop life.

But, under the conditions that actually prevail, the very reverse is the case. The quantity of matter remaining between our cluster and the Milky Way being comparatively small, the aggregation into suns has gone on more regularly and more slowly. The motions acquired by our sun and its neighbours have been rendered moderate by two causes: (1) their nearness to the centre of the very slowly aggregating cluster where the motion due to gravitation is least in amount; and (2) the slight differential attraction away from the centre by the Milky Way on the side nearest to us. Again, this protective action of the Milky Way has been repeated, on a smaller scale, by the formation of the outer ring of the solar cluster, which has thus preserved the inner central cluster itself from a too abundant direct inflow of large masses of matter.

But although the matter composing the outer portion of the original universe has been to a large extent aggregated into the vast system of the Milky Way, it seems probable, perhaps even certain, that some portion would escape its attractive forces and would pass through its numerous open spaces—indicated by the dark rifts, channels, and patches, as already described—and thus flow on unchecked towards the centre of mass of the whole system. The quantity of matter thus reaching the central cluster from the enormously remote spaces beyond the Milky Way might be very small in comparison with what was retained to build up that wonderful star-system; but it might yet be so large in total amount as to play an important part in the formation of the central group of suns. It would probably flow inwards almost continuously, and when it ultimately reached the solar cluster, it would have attained a very high velocity. If, therefore, it were widely diffused, and consisted of masses of small or moderate size as compared with planets or stars, it would furnish the energy requisite for bringing these slowly aggregating stars to the required intensity of heat for forming luminous suns.

Here, then, I think, we have found an adequate

explanation of the very long-continued light and heat-emitting capacity of our sun, and probably of many others in about the same position in the solar cluster. These would at first gradually aggregate into considerable masses from the slowly moving diffused matter of the central portions of the original universe; but at a later period they would be reinforced by a constant and steady inrush of matter from its very outer regions, and therefore possessing such high velocities as to materially aid in producing and maintaining the requisite temperature of a sun such as ours, during the long periods demanded for continuous life-development. The enormous extension and mass of the original universe of diffused matter (as postulated by Lord Kelvin) is thus seen to be of the greatest importance as regards this ultimate product of evolution, because, without it, the comparatively slow-moving and cool central regions might not have been able to produce and maintain the requisite energy in the form of heat; while the aggregation of by far the larger portion of its matter in the great revolving ring of the galaxy was equally important, in order to prevent the too great and too rapid inflow of matter to these favoured regions.

It appears, then, that if we admit as probable some such process of development as I have here indicated, we can dimly see the bearing of all the great features of the stellar universe upon the successful development of life. These are, its vast dimensions; the form it has acquired in the mighty ring of the Milky Way; and our position near to, but not exactly in, its centre. We know that the star-system *has* acquired these forms, presumably from some simple and more diffused condition. We know that we *are* situated near the centre of this vast system. We know that our sun *has* emitted light and heat, almost uniformly, for periods incompatible with rapid aggregation and the equally rapid cooling which physicists consider inevitable. I have here suggested a mode of development which would lead to a very slow but continuous growth of the more central suns; to an excessively long period of nearly stationary heat-giving power; and lastly, an equally long period of very gradual cooling—a period the commencement of which our sun may have just entered upon.

Descending now to terrestrial physics, I have shown that, owing to the highly complex nature of the adjustments required to render a world habitable and to retain its habitability during the æons of time requisite for life-development, it is in the highest degree improbable that the required conditions and adaptations should have occurred in any other planets of any other suns, which *might* occupy an equally favourable position with our own, and which were of the requisite size and heat-giving power.

Lastly, I submit that the whole of the evidence I have here brought together leads to the conclusion that our earth is almost certainly the only inhabited planet in our solar system; and, further, that there is no inconceivability—no improbability even—in the conception that, in order to produce a world that should be precisely adapted in every detail for the orderly development of organic life culminating in man, such a vast and complex universe as that which we know exists around us, may have been absolutely required.

Summary of Argument

As the last ten chapters of this volume embody a connected argument leading to the conclusion above stated, it may be useful to my readers to summarise rather fully the successive steps of this argument, the facts on which it rests, and the various subsidiary conclusions arrived at.

(1) One of the most important results of modern astronomy is to have established the unity of the vast stellar universe which we see around us. This rests upon a great mass of observations, which demonstrate the wonderful complexity in detail of the arrangement and distribution of stars and nebulæ, combined with a no less remarkable general symmetry, indicating throughout a single inter-dependent system, not a number of totally distinct systems so far apart as to have no physical relations with each other, as was once supposed.

(2) This view is supported by numerous converging

lines of evidence, all tending to show that the stars are not infinite in number, as was once generally believed, and which view is even now advocated by some astronomers. The very remarkable calculations of Lord Kelvin, referred to in the early part of this chapter, give a further support to this view, since they show that if the stars extended much beyond those we see or can obtain direct knowledge of, and with no very great change in their average distance apart, then the force of gravitation towards the centre would have produced on the average more rapid motions than the stars generally possess.

(3) An overwhelming consensus of opinion among the best astronomers establishes the fact of our nearly central position in the stellar universe. They all agree that the Milky Way is nearly circular in form. They all agree that our sun is situated almost exactly in its medial plane. They all agree that our sun, although not situated at the exact centre of the galactic circle, is yet not very far from it, because there are no unmistakable signs of our being nearer to it at any one point and farther away from the opposite point. Thus the nearly central position of our sun in the great star-system is almost universally admitted.

On the question of the solar-cluster there is more difference of opinion; though here, again, all are agreed that there is such a cluster. Its size, form, density, and exact position are somewhat uncertain, but I have, as far as possible, been guided by the best available evidence. If we adopt Lord Kelvin's general idea of the gradual condensation of an enormous diffused mass of matter towards its common centre of gravity, that centre would be approximately the centre of this cluster. Also, as gravitational force at and near this centre would be comparatively small, the motions produced there would be slow, and collisions, being due only to differential motions, when they did occur would be very gentle. We might therefore expect many dark aggregations of matter here, which may explain why we do not find any special crowding of visible stars in the direction of this centre; while, as no star has a sensible disc, the dark stars if at great distances would hardly ever be seen to occult the bright ones. Thus, it seems to me, the controlling force may be explained which has retained our sun in approximately the same orbit around the centre of gravity of this central cluster during the whole period of its existence as a sun and our existence as a planet; and has thus saved us from the possibility—perhaps even the certainty—of disastrous collisions or disruptive approaches to which suns, in or near the Milky Way, and to a less extent elsewhere, are or have been exposed. It seems quite probable that in that region of more rapid and less controlled motions and more crowded masses of matter, no star can remain in a nearly stable condition as regards temperature for sufficiently long periods to allow of a complete system of life-development on any planet it may possess.

(4) The various proofs are next stated that assure us of the almost complete uniformity of matter, and of material physical and chemical laws, throughout our universe. This I believe no one seriously disputes; and it is a point of the greatest importance when we come to consider the conditions required for the development and maintenance of life, since it assures us that very similar, if not identical, conditions must prevail wherever organic life is or can be developed.

(5) This leads us on to the consideration of the essential characteristics of the living organism, consisting as it does of some of the most abundant and most widely distributed of these material elements, and being always subject to the general laws of matter. The best authorities in physiology are quoted, as to the extreme complexity of the chemical compounds which constitute the physical basis for the manifestation of life; as to their great instability; their wonderful mobility combined with permanence of form and structure; and the altogether marvellous powers they possess of bringing about unique chemical transformations and of building up the most complicated structures from simple elements.

I have endeavoured to put the broad phenomena of vegetable and animal life in a way that will enable my readers to form some faint conception of the intricacy, the delicacy, and the mystery of the myriad living forms they see everywhere around them. Such a conception will enable them to realise how supremely grand is organic life, and to appreciate better, perhaps, the absolute necessity for the numerous, complex and delicate adaptations of inorganic nature, without which it is impossible for life either to exist now, or to have been developed during the immeasurable past.

(6) The general conditions which are absolutely essential for life thus manifested on our planet are then discussed, such as, solar light and heat; water universally distributed on the planet's surface and in the atmosphere; an atmosphere of sufficient density, and composed of the several gases from which alone protoplasm can be formed; some alternations of light and darkness, and a few others.

(7) Having treated these conditions broadly, and explained why they are important and even indispensable for life, we next proceed to show how they are fulfilled upon the earth, and how numerous, how complex, and often how exact are the adjustments needed to bring them about, and maintain them almost unchanged throughout the vast æons of time occupied in the development of life. Two chapters are devoted to this subject; and it is believed that they contain facts that will be new to many of my readers. The combinations of causes which lead to this result are so varied, and in several cases dependent on such exceptional peculiarities of physical constitution, that it seems in the highest degree improbable

that they can *all* be found again combined either in the solar system or even in the stellar universe. It will be well here just to enumerate these conditions, which are all essential within more or less narrow limits:—

Distance of planet from the sun.

Mass of planet.

Obliquity of its ecliptic.

Amount of water as compared with land.

Surface distribution of land and water.

Permanence of this distribution, dependent probably on the unique origin of our moon.

An atmosphere of sufficient density, and of suitable component gases.

An adequate amount of dust in the atmosphere.

Atmospheric electricity.

Many of these act and react on each other, and lead to results of great complexity.

(8) Passing on to other planets of the solar system, it is shown that none of them combine all the complex conditions which are found to work harmoniously together on the earth; while in most cases there is some one defect which alone removes them from the category of possible life-producing and life-supporting planets. Among these are the small size and mass of Mars, being such that it cannot

retain aqueous vapour; and the fact that Venus rotates on its axis in the same time as it takes to revolve round the sun. Neither of these facts was known when Proctor wrote upon the question of the habitability of the planets. All the other planets are now given up—and were given up by Proctor himself—as possible life-bearers in their present stage; but he and others have held that, if not suitable now, some of them may have been the scene of life-development in the past, while others will be so in the future.

In order to show the futility of this supposition, the problem of the duration of the sun as a stable heat-giver is discussed; and it is shown that it is only by reducing the periods claimed by geologists and biologists for life-development upon the earth, and by extending the time allowed by physicists to its utmost limits, that the two claims can be harmonised. It follows that the whole period of the sun's duration as a light and heat-giver has been required for the development of life upon the earth; and that it is only upon planets whose phases of development synchronise with that of the earth that the evolution of life is possible. For those whose material evolution has gone on quicker or slower, there has not been, or will not be, time enough for the development of life.

(9) The problem of the stars as possibly having life-supporting planets is next dealt with, and reasons are given why in only a minute portion of the whole is this possible.

Even in that minute portion, reduced probably to a few of the component suns of the solar-cluster, a large proportion seems likely to be ruled out by being close binary systems, and another large portion by being in process of aggregation. In those remaining, whether they may be reckoned by tens or by hundreds we cannot say, the chances against the same complex combination of conditions as those which we find on the earth occurring on any planet of any other sun are enormously great.

(10) I then refer, briefly, to some recent measurements of star-radiation, and suggest that they may thus possibly have important effects on the development of vegetable and animal life; and, finally, I discuss the problem of the stability of the stellar universe and the special advantage we derive from our central position, suggested by some of the latest researches of our great mathematician and physicist—Lord Kelvin.

Conclusions

Having thus brought together the whole of the available evidence bearing upon the questions treated in this volume, I claim that certain definite conclusions have been reached and proved, and that certain other conclusions have enormous probabilities in their favour.

The conclusions reached by modern astronomers are: (1) That the stellar universe forms one connected whole; and, though of enormous extent, is yet finite, and its extent determinable.

(2) That the solar system is situated in the plane of the Milky Way, and not far removed from the centre of that plane. The earth is therefore nearly in the centre of the stellar universe.

(3) That this universe consists throughout of the same kinds of matter, and is subjected to the same physical and chemical laws.

The conclusions which I claim to have shown to have enormous probabilities in their favour are—

(4) That no other planet in the solar system than our earth is inhabited or habitable.

(5) That the probabilities are almost as great against any other sun possessing inhabited planets.

(6) That the nearly central position of our sun is probably a permanent one, and has been specially favourable, perhaps absolutely essential, to life-development on the earth.

These latter conclusions depend upon the combination of a large number of special conditions, each of which must be in definite relation to many of the others, and must all have persisted simultaneously during enormous periods of time. The weight to be given to this kind of reasoning depends upon a full and fair consideration of the *whole* evidence as I have endeavoured to present it in the last seven chapters of this book. To this evidence I appeal.

This completes my work as a connected argument, founded wholly on the facts and principles accumulated by modern science; and it leads, if my facts are substantially correct and my reasoning sound, to one great and definite conclusion—that man, the culmination of conscious organic life, has been developed here only in the whole vast material universe we see around us. I claim that this is the logical outcome of the evidence, if we consider and weigh this evidence without any prepossessions whatever. I maintain that it is a question as to which we have no right to form *a priori* opinions not founded upon evidence. And evidence opposed to this conclusion, or even as to its improbability, we have absolutely none whatever.

But, if we admit the conclusion, nothing that need alarm either the scientific or the religious mind necessarily follows, because it can be explained or accounted for in either of two distinct ways. One considerable body, including probably the majority of men of science, will admit that the evidence does apparently lead to this conclusion, but will explain it as due to a fortunate coincidence. There might have been a hundred or a thousand life-bearing planets, had the course of evolution of the universe been a little different, or there might have been none at all. They would probably add, that, as life and man *have* been produced, that shows that their production was possible; and therefore, if not now then at some other time, if not here then in some other planet of some other sun, we should be sure to have come into existence; or if not precisely the same as we are, then something a little better or a little worse.

The other body, and probably much the largest, would be represented by those who, holding that mind is essentially superior to matter and distinct from it, cannot believe that life, consciousness, mind, are products of matter. They hold that the marvellous complexity of forces which appear to control matter, if not actually to constitute it, are and must be mind-products; and when they see life and mind apparently rising out of matter and giving to its myriad forms an added complexity and unfathomable mystery, they see in this development an additional proof of the supremacy of mind. Such persons would be inclined to the belief of the great eighteenth century scholar, Dr. Bentley, that the soul of one virtuous man is of greater worth and excellency than the sun and all his planets and all the stars in the heavens; and when they are shown that there are strong reasons for thinking that man *is* the unique and supreme product of this vast universe, they will see no difficulty in going a little further, and believing that the universe was actually brought into existence for this very purpose.

With infinite space around us and infinite time before and behind us, there is no incongruity in this conception. A universe as large as ours for the purpose of bringing into existence many myriads of living, intellectual, moral, and spiritual beings, with unlimited possibilities of life and happiness, is surely not *more* out of proportion than is the complex machinery, the life-long labour, the ingenuity and invention which we have bestowed upon the production of the humble, the trivial, *pin*. Neither is the apparent waste of energy so great in such a universe, comparatively, as the millions of acorns, produced during its life by an oak, every one of which might grow to be a tree, but of which only *one* does actually, after several hundred years, produce the *one* tree which is to replace the parent. And if it is said that the acorns are food for bird and beast, yet the spores of ferns and the seeds of orchids are not so, and countless millions of these go to waste for every one which reproduces the parent form. And all through the animal world, especially among the lower types, the same thing is seen. For the great majority of these entities *we* can see no use whatever, either of the enormous variety of the species, or the vast hordes of individuals. Of beetles alone there are at least a hundred thousand distinct species now living, while in some parts of sub-arctic America mosquitoes are sometimes so excessively abundant that they obscure the sun. And when we think of the myriads that have existed through the vast ages of geological time, the mind reels under the immensity of, to us, apparently useless life.

All nature tells us the same strange, mysterious story, of the exuberance of life, of endless variety, of unimaginable quantity. All this life upon our earth has led up to and culminated in that of man. It has been, I believe, a common and not unpopular idea that during the whole process of the rise and growth and extinction of past forms, the earth has been preparing for the ultimate—Man. Much of the wealth and luxuriance of living things, the infinite variety of form and structure, the exquisite grace and beauty in bird and insect, in foliage and flower, may have been mere by-products of the grand mechanism we call nature—the one and only method of developing humanity.

And is it not in perfect harmony with this grandeur of design (if it be design), this vastness of scale, this marvellous process of development through all the ages, that the material universe needed to produce this cradle of organic life, and of a being destined to a higher and a permanent existence, should be on a corresponding scale of vastness, of complexity, of beauty? Even if there were no such evidence as I have here adduced for the unique position and the exceptional characteristics which distinguish the earth, the old idea that all the planets were inhabited, and that all the stars existed for the sake of other planets, which planets existed to develop life, would, in the light of our present knowledge, seem utterly improbable and incredible. It would introduce monotony into a universe whose grand character and teaching is endless diversity. It would imply that to produce the living soul in the marvellous and glorious body of man—man with his faculties, his aspirations, his powers for good and evil—that this was an easy matter which could be brought about anywhere, in any world. It would imply that man is an animal and nothing more, is of no importance in the universe, needed no great preparations for his advent, only, perhaps, a second-rate demon, and a third or fourth-rate earth. Looking at the long and slow and complex growth of nature that preceded his appearance, the immensity of the stellar universe with its thousand million suns, and the vast æons of time during which it has been developing—all these seem only the appropriate and harmonious surroundings, the necessary supply of material, the sufficiently spacious workshop for the production of that planet which was to produce first, the organic world, and then, Man.

In one of his finest passages our great world-poet gives us *his* conception of the grandeur of human nature—'What a piece of work is man! How noble in reason! How infinite in faculty! In form and moving, how express and admirable! In action how like an angel! In apprehension how like a god!' And for the development of such a being what is a universe such as ours? However vast it may seem to our faculties, it is as a mere nothing in the ocean of the infinite. In infinite space there may be infinite universes, but I hardly think they would be all universes of matter. That would indeed be a low conception of infinite power! Here, on earth, we see millions of distinct species of animals, millions of different species of plants, and each and every species consisting often of many millions of individuals, no two individuals exactly alike; and when we turn to the heavens, no two planets, no two satellites alike; and outside our system we see the same law prevailing—no two stars, no two clusters, no two nebulæ alike. Why then should there be other universes of the *same* matter and subject to the *same* laws—as is implied by the conception that the stars are infinite in number, and extend through infinite space?

Of course there may be, and probably are, other universes, perhaps of other kinds of matter and subject to other laws, perhaps more like our conceptions of the ether, perhaps wholly non-material, and what we can only conceive of as spiritual. But, unless these universes, even though each of them were a million times vaster than our stellar universe, were also infinite in number, they could not fill infinite space, which would extend on all sides beyond them, so that even a million million such universes would shrink to imperceptibility when compared with the vast beyond!

Of infinity in any of its aspects we can really know nothing, but that it exists and is inconceivable. It is a thought that oppresses and overwhelms. Yet many speak of it glibly as if they *knew* what it contains, and even use that assumed knowledge as an argument against views that are unacceptable to themselves. To me its existence is absolute but unthinkable—that way madness lies.

'O night! O stars, too rudely jars The finite with the infinite!'

I will conclude with one of the finest passages relating to the infinite that I am acquainted with, from the pen of the late R.A. Proctor:

'Inconceivable, doubtless, are these infinities of time and space, of matter, of motion, and of life. Inconceivable that the whole universe can be for all time the scene of the operation of infinite power, omnipresent, all-knowing. Utterly incomprehensible how Infinite Purpose can be associated with endless material evolution. But it is no new thought, no modern discovery, that we are thus utterly powerless to conceive or comprehend the idea of an Infinite Being, Almighty, All-knowing, Omnipresent, and Eternal, of whose inscrutable purpose the material universe is the unexplained manifestation. Science is in presence of the old, old mystery; the old, old questions are asked of her—"Canst thou by searching find out God? Canst thou find out the Almighty unto perfection? It is as high as heaven; what canst thou do? deeper than hell; what canst thou know?" And science answers these questions as they were answered of old—"As touching the Almighty we cannot find Him out.'"

The following beautiful lines—among the latest products of Tennyson's genius—so completely harmonise with the subject-matter of the present volume, that no apology is needed for quoting them here:—

(*The Question*)

Will my tiny spark of being Wholly vanish in your deeps and heights? Must my day be dark by reason, O ye Heavens, of your boundless nights, Rush of Suns and roll of systems, And your fiery clash of meteorites?

(*The Answer*)

'Spirit, nearing yon dark portal At the limit of thy human state, Fear not thou the hidden purpose Of that Power which alone is great, Nor the myriad world, His shadow, Nor the silent Opener of the Gate.'

I

THE NEBULÆ AND CLUSTERS OF THE
NORTHERN HEAVENS.

Drawn upon an equal surface projection from Dr.
DREYER'S catalogue of 1888.
The Milky Way from a Dr. BOEDDICKER'S Drawing.
By SIDNEY WATERS

THE NEBULÆ AND CLUSTERS OF THE SOUTHERN HEAVENS.

Drawn upon an equal surface projection from Dr. DREYER'S catalogue of 1888.
The Milky Way from URANOMETRIA ARGENTINA.
By SIDNEY WATERS

FOOTNOTES:

In a letter to *Knowledge*, June 1903, Mr. W.H.T. Monck puts the same point in a mathematical form.

Outlines of Astronomy, pp. 578-9. In the passages quoted the italics are Sir John Herschel's.

Mr. J.E. Gore in *Concise Knowledge Astronomy*, pp. 541-2.

Nature, October 26, 1899.

The System of the Stars, p. 385.

This account of Professor Kapteyn's research is taken from an article by Miss A.M. Clerke in *Knowledge*, April 1893.

The Stars, p. 256. The region here referred to is that where the Milky Way has its greatest width (though nearly as wide in the part exactly opposite), and where it may perhaps extend somewhat in our direction. Miss A.M. Clerke informs me that in April 1901 Kapteyn withdrew the conclusions arrived at in 1893, as being founded on illegitimate reasoning as to the relation of parallaxes to proper motions. But as this relation is still accepted, under certain limitations, by Professor Newcomb and other astronomers, who have arrived independently at very similar results, it seems not improbable that, after all, Professor Kapteyn's conclusions may not require very much modification. Professor Newcomb also tells us (*The Stars*, p. 214, footnote) that he has seen the latest of Professor Kapteyn's papers, down to 1901; but he does not therefore express any doubt as to his own conclusions as here referred to.

See *Knowledge* and *The Fortnightly Review* of April 1903.

Sir R. Ball in an article in *Good Words* (April 1903) says that luminosity is an exceptional phenomenon in nature, and that luminous stars are but the glow-worms and fireflies of the universe, as compared with the myriads of other animals.

The Astrophysical Journal, vol. xiv., July 1901, p. 17.

Professor F.J. Allen: *What is Life?*

Art. 'Vegetable Physiology' in *Chambers's Encyclopædia*.

Address to the British Association1902, Section Physiology.

This enumeration of the elements that enter into the structure of plants and animals is taken from Professor F.J. Allen's paper already referred to.

Early History of Mankind 2nd ed. p. 227.

For a fuller account of this Arctic fauna and flora see the works of Sir C. Lyell, Sir A. Geikie, and other geologists. A full summary of it is also given in the author's *Island Life*.

Professor G.H. Darwin states that it is nearly certain that no other satellite nor any of the planets originated in the same way as the moon.

Transactions of Royal Dublin Society, vol. vi. (ser. ii.), part xiii. 'Of Atmospheres upon Planets and Satellites.' By G. Johnstone Stoney, F.R.S., etc. etc.

Knowledge, June 1903.

M. Camille Flammarion, in *Knowledge*, June 1903.

The Worlds of Space, by J.E. Gore, chapter iii.

The Fortnightly Review, April 1903, p. 60.

Since writing this chapter I have seen a paper by Luigi d'Auria dealing mathematically with 'Stellar Motion, etc.,' and am pleased to see that, from quite different considerations, he has found it necessary to place the solar system at a distance from the centre not very much more remote than the position I have given it. He says: 'We have good reasons to suppose that the solar system is rather near the centre of the Milky Way, and as this centre would, according to our hypothesis, coincide with the centre of the Universe, the distance of 159 light years assumed is not too great, nor can it be very much smaller.'—*Journal of the Franklin Institute*, March 1903.

WORKS BY THE SAME AUTHOR.

Travels on the Amazon and Rio-Negro.

Palm-trees of the Amazon, and their Uses (*out of print*).

The Malay Archipelago.

Natural Selection and Tropical Nature.

The Geographical Distribution of Animals. *2 vols.*

Island Life, or Insular Faunas and Floras.

Darwinism.

Australasia. *2 vols.*

Bad Times—Causes of Depression of Trade (*out of print*).

Land Nationalisation, its Necessity and its Aims.

Vaccination a Delusion.

Miracles and Modern Spiritualism.

The Wonderful Century (*New and Illustrated Edition*).

The Wonderful Century Reader.

Studies Scientific and Social.